UAS
Pilot Log

Unmanned Aircraft Systems
Logbook for Drone Pilots & Operators

created by
droneprep
droneprep.com
@droneprep

UAS Pilot Log: Unmanned Aircraft Systems Logbook for Drone Pilots and Operators

By publishing this book, neither the authors nor the publisher are engaged in rendering legal or other professional services. If any such assistance is required, the services of a qualified professional or should be sought. The authors and publisher will not be responsible for any liability, loss, or risk incurred as a result of the use and application of any of the information contained in this book.

All rights reserved. No portion of this book may be reproduced, stored in a retrieval system, or transmitted in any form or by any means—electronic, mechanical, photocopy, recording, scanning, or other, without the prior written permission of the publisher and/or authors.

Copyright © 2015 droneprep.com
All rights reserved.

UAS Pilot Log: Unmanned Aircraft Systems Logbook for Drone Pilots and Operators

Logbook Onwer Information

First and Last Name:

Contact Information:

    E-mail Address:

    Physical Address:

    Phone:

Certificates & Ratings:

Aircrafts Owned:

How To Use This Log Book

Congratulations on your decision to purchase Unmanned Aircraft Systems: Drone Pilot and Operator Log. After extensive research into record keeping and processes maintained by expert operators, droneprep.com designed this log book specifically for pilots and operators of unmanned aircraft systems and drones.

This logbook distills complex procedures and note taking with simple, easy-to-understand entry pages that can be maintained by any drone operator, regardless of skill level or experience. The result is a flexible and powerful record that will serve as both a tool to enhance your flying experience and a superb record of exactly what happened on the day of your flight.

To help you get started, what follows is a brief tutorial that explains how to use each section of the log book so you can maximize your flying experience.

1. Flight Session Information

The Flight Session Information section gives you an opportunity to describe each flying session. Since a given flying session may involve multiple takeoffs and landings, we describe below how to reflect the flying time for each interval below under the "Flight Notes") section. We therefore recommend you identify the "Flight Session" with a particular identification or number so it is unique.

FLIGHT ID / NO.: _____   Month _____ Day _____ Year _____

FLIGHT LOCATION: _____   Weather: _____

For example, under **Flight ID / NO.**, we suggest establishing your own numbering system so you can differentiate one flying session from the next. This will help you maintain a process driven approach to record keeping that is both standardized and applicable to future flights.

Under **Flight Location** you can be as detailed as you like. The goal is to simply memorialize the starting point for your flight along with the **Date** and **Weather** conditions. We recommend populating every field as these details become useful later as you start to analyze data across multiple flights.

2. Aircraft and Crew Data

The Aircraft and Crew Data section is straightforward and designed to capture details about the **Unmanned Aircraft System** used in a flight session's operations. We suggest this to be as detailed as necessary to capture the identity of craft so there is a reasonable level of uniqueness recorded, including the craft's **manufacturer** and **model number**.

UNMANNED AIRCRAFT SYSTEM: **CREW:**

MANUFACTURER: _____   PILOT: _____

MODEL NUMBER: _____   SPOTTER: _____

This section also captures personnel who are involved in a given flight session. The requirement to detail the identity of both the **Pilot in Command** and **Spotter** highlights the importance of multi-person approaches to flying. The use of a spotter during flight operations lends itself to greater safety and more successful operations.

3. Preflight Checklist

Every Pilot in Command should make preflight checks and procedures part of the regular routine before each flight. While every pilot may approach preflight checks differently, we provide an opportunity for you to confirm that some of the most common aircraft checks are completed. Here is a sample of the section:

PREFLIGHT CHECKLIST:
- ☐ Batteries Charged & Secure
- ☐ Aircraft Hardware OK
- ☐ Equipment & Gear OK
- ☐ Transmitter Controls OK
- ☐ Props OK & Tight
- ☐ Software / Firmware Update
- ☐ Transmitter Control Power ON
- ☐ Aircraft Power ON
- ☐ Compass Calibration
- ☐ Camera / FPV ON
- ☐ Satellite Connection
- ☐ Applications / Systems ON

PREFLIGHT NOTES:

As shown, you also have an area to detail any **Preflight Notes** to memorialize any activities or circumstances that may relevant to the day's flight or safety operations. Please use the above checks as a guide and not necessarily an exhaustive list. Remember, the goal is to make sure you, your crew, your craft and the people around you are safe at all times. Preflight are one way you can achieve this, but ultimately it's up to you to make sure

4. Session Flight Intervals

As a UAS or drone operator, the reality is that you may take your craft up multiple times in a single day's session. This may be due to a handful of reasons, including battery life, video recording space or those unfortunate premature and unplanned landings. Unlike other log books, which are based on traditional airplane style operations, we designed this log to reflect the practical realities of the UAS and drone pilot who operate small crafts.

To populate the **Session Flight Intervals** table, start by entering details on line number 1 for your first flight within a given session. Adjacent to the number 1 (or numbers 2 through 7), you may enter a name, number or any other unique identifier to represent that specific interval within the session.

For **START**, enter either the time of day or, if using a stopwatch method of timing, the starting time. For **STOP**, enter the corresponding time of day or stopping time point. Repeat the same log entry practice for up to seven intervals. Should you need more intervals for the same flight session, simply use another page.

SESSION FLIGHT INTERVALS	FLIGHT TIMES		
	START	STOP	TOTAL
1.			
2.			
3.			
4.			
5.			
6.			
7.			
TOTAL HOURS FOR SESSION			
TOTAL FORWARD			
TOTAL TO DATE			

Once your intervals and your flight session is completed for the day, sum the TOTAL lines together and enter the value in the **TOTAL HOURS FOR SESSION** box. Underneath the TOTAL HOURS FOR SESSION box is **TOTAL FORWARD**, which is the **TOTAL TO DATE** value from the immediately prior flight session. If you are making your first very first flight session entry, there is no TOTAL FORWARD. **TOTAL TO DATE** is therefore the sum of TOTAL HOURS FOR SESSION plus the TOTAL FORWARD.

5. Flight Map

The **Flight Map** affords the operator a chance to visually depict or draw various elements from the flight session. This section is optional and should be flexible to the needs of the operator, but we have found it can add value from a preflight planning or post flight analysis standpoint. For example, you may want to plot your flight path and any relevant way points. Alternatively, you may also want to use this area to help you conceptualize the flying field and any possible obstructions or elevation changes. Ultimately, you may use this area as you see fit to promote a safe and effective flight and to capture a record of exactly what happened during your flight session.

6. Other Entries

The remaining entries simply allow you to enter any **Post flight Notes** and **Journal Entries**. After your flight, take a few minutes to reflect on what happened and whether any of the session's events are worth recording for future reference. Once you have completed your entries, please sign the bottom of the page—along with the Spotter—to certify that your entries are true and correct.

UAS Pilot Log: Unmanned Aircraft Systems Logbook for Drone Pilots and Operators

Log Entry Pages

droneprep.com

UAS Pilot Log: Unmanned Aircraft Systems Logbook for Drone Pilots and Operators

FLIGHT ID / NO.: **PRAC 01**
FLIGHT LOCATION: **MORTON FARM DARLINGTON**

Month **11** Day **11** Year **2016**
Weather: **CLEAR**

UNMANNED AIRCRAFT SYSTEM:

MANUFACTURER: **DJI**
MODEL NUMBER: **PHANTOM 3 ADVANCED**

CREW:

PILOT: **JESSADA STEVENSON**
SPOTTER: ―

PREFLIGHT CHECKLIST:

- ☑ Batteries Charged & Secure
- ☑ Aircraft Hardware OK
- ☑ Equipment & Gear OK
- ☑ Transmitter Controls OK
- ☑ Props OK & Tight
- ☑ Software / Firmware Update
- ☑ Transmitter Control Power ON
- ☑ Aircraft Power ON
- ☑ Compass Calibration
- ☑ Camera / FPV ON
- ☑ Satellite Connection
- ☑ Applications / Systems ON

PREFLIGHT NOTES:

DURHAM TEES AIRPORT NOTIFIED AT 14:25.

SESSION FLIGHT INTERVALS	FLIGHT TIMES START	FLIGHT TIMES STOP	FLIGHT TIMES TOTAL
1.	15:30	16:00	
2.			
3.			
4.			
5.			
6.			
7.			
TOTAL HOURS FOR SESSION			0.5
TOTAL FORWARD			0.5
TOTAL TO DATE			0.5

FLIGHT MAP

POSTFLIGHT NOTES / JOURNAL ENTRIES:

I CERTIFY THAT THE FOREGOING ENTRIES ARE TRUE AND CORRECT:

PILOT: **JESSADA STEVENSON**   SPOTTER:

UAS Pilot Log: Unmanned Aircraft Systems Logbook for Drone Pilots and Operators

FLIGHT ID / NO.: PRAC02
FLIGHT LOCATION: MORTON FARM.

Month 12 Day 11 Year 2016
Weather: _____

UNMANNED AIRCRAFT SYSTEM:

MANUFACTURER: DJI
MODEL NUMBER: PHANTOM 3

CREW:

PILOT: JESSADA STEVENSON
SPOTTER: AMY STEVENSON

PREFLIGHT CHECKLIST:

- [x] Batteries Charged & Secure
- [x] Aircraft Hardware OK
- [x] Equipment & Gear OK
- [x] Transmitter Controls OK
- [x] Props OK & Tight
- [x] Software / Firmware Update
- [x] Transmitter Control Power ON
- [x] Aircraft Power ON
- [x] Compass Calibration
- [x] Camera / FPV ON
- [ ] Satellite Connection
- [x] Applications / Systems ON

PREFLIGHT NOTES:

Air Traffic notified at 12:59.

SESSION FLIGHT INTERVALS	FLIGHT TIMES		
	START	STOP	TOTAL
1.	16:3		
2.	15:40	16:02	0.5
3.			
4.			
5.			
6.			
7.			
TOTAL HOURS FOR SESSION			0.5
TOTAL FORWARD			0.5
TOTAL TO DATE			1 HR

FLIGHT MAP

grain tracks

takeoff X
Road
Bridge

POSTFLIGHT NOTES / JOURNAL ENTRIES:

I CERTIFY THAT THE FOREGOING ENTRIES ARE TRUE AND CORRECT:

PILOT: _____
SPOTTER: _____

droneprep.com

UAS Pilot Log: Unmanned Aircraft Systems Logbook for Drone Pilots and Operators

FLIGHT ID / NO.: _____   Month _____ Day _____ Year _____

FLIGHT LOCATION: _____   Weather: _____

UNMANNED AIRCRAFT SYSTEM: **CREW:**

MANUFACTURER: _____   PILOT: _____

MODEL NUMBER: _____   SPOTTER: _____

PREFLIGHT CHECKLIST:
- ☐ Batteries Charged & Secure
- ☐ Aircraft Hardware OK
- ☐ Equipment & Gear OK
- ☐ Transmitter Controls OK
- ☐ Props OK & Tight
- ☐ Software / Firmware Update
- ☐ Transmitter Control Power ON
- ☐ Aircraft Power ON
- ☐ Compass Calibration
- ☐ Camera / FPV ON
- ☐ Satellite Connection
- ☐ Applications / Systems ON

PREFLIGHT NOTES:

SESSION FLIGHT INTERVALS	FLIGHT TIMES		
	START	STOP	TOTAL
1.			
2.			
3.			
4.			
5.			
6.			
7.			
TOTAL HOURS FOR SESSION			
TOTAL FORWARD			
TOTAL TO DATE			

FLIGHT MAP

POSTFLIGHT NOTES / JOURNAL ENTRIES:

I CERTIFY THAT THE FOREGOING ENTRIES ARE TRUE AND CORRECT:

PILOT: _____   SPOTTER: _____

droneprep.com

UAS Pilot Log: Unmanned Aircraft Systems Logbook for Drone Pilots and Operators

FLIGHT ID / NO.: _____   Month _____ Day _____ Year _____

FLIGHT LOCATION: _____   Weather: _____

UNMANNED AIRCRAFT SYSTEM: **CREW:**

MANUFACTURER: _____   PILOT: _____

MODEL NUMBER: _____   SPOTTER: _____

PREFLIGHT CHECKLIST:

- ☐ Batteries Charged & Secure
- ☐ Aircraft Hardware OK
- ☐ Equipment & Gear OK
- ☐ Transmitter Controls OK
- ☐ Props OK & Tight
- ☐ Software / Firmware Update
- ☐ Transmitter Control Power ON
- ☐ Aircraft Power ON
- ☐ Compass Calibration
- ☐ Camera / FPV ON
- ☐ Satellite Connection
- ☐ Applications / Systems ON

PREFLIGHT NOTES:

_____

SESSION FLIGHT INTERVALS	FLIGHT TIMES		
	START	STOP	TOTAL
1.			
2.			
3.			
4.			
5.			
6.			
7.			
TOTAL HOURS FOR SESSION			
TOTAL FORWARD			
TOTAL TO DATE			

FLIGHT MAP

POSTFLIGHT NOTES / JOURNAL ENTRIES:

I CERTIFY THAT THE FOREGOING ENTRIES ARE TRUE AND CORRECT:

PILOT: _____   SPOTTER: _____

droneprep.com

UAS Pilot Log: Unmanned Aircraft Systems Logbook for Drone Pilots and Operators

FLIGHT ID / NO.: _____

Month _____ Day _____ Year _____

FLIGHT LOCATION: _____

Weather: _____

UNMANNED AIRCRAFT SYSTEM:

MANUFACTURER: _____

MODEL NUMBER: _____

CREW:

PILOT: _____

SPOTTER: _____

PREFLIGHT CHECKLIST:

- ☐ Batteries Charged & Secure
- ☐ Aircraft Hardware OK
- ☐ Equipment & Gear OK
- ☐ Transmitter Controls OK
- ☐ Props OK & Tight
- ☐ Software / Firmware Update
- ☐ Transmitter Control Power ON
- ☐ Aircraft Power ON
- ☐ Compass Calibration
- ☐ Camera / FPV ON
- ☐ Satellite Connection
- ☐ Applications / Systems ON

PREFLIGHT NOTES:

_____

SESSION FLIGHT INTERVALS	FLIGHT TIMES		
	START	STOP	TOTAL
1.			
2.			
3.			
4.			
5.			
6.			
7.			
TOTAL HOURS FOR SESSION			
TOTAL FORWARD			
TOTAL TO DATE			

FLIGHT MAP

POSTFLIGHT NOTES / JOURNAL ENTRIES:

I CERTIFY THAT THE FOREGOING ENTRIES ARE TRUE AND CORRECT:

PILOT: _____   SPOTTER: _____

UAS Pilot Log: Unmanned Aircraft Systems Logbook for Drone Pilots and Operators

FLIGHT ID / NO.: _____   Month _____ Day _____ Year _____

FLIGHT LOCATION: _____   Weather: _____

UNMANNED AIRCRAFT SYSTEM: **CREW:**

MANUFACTURER: _____   PILOT: _____

MODEL NUMBER: _____   SPOTTER: _____

PREFLIGHT CHECKLIST:

- ☐ Batteries Charged & Secure   ☐ Props OK & Tight   ☐ Compass Calibration
- ☐ Aircraft Hardware OK   ☐ Software / Firmware Update   ☐ Camera / FPV ON
- ☐ Equipment & Gear OK   ☐ Transmitter Control Power ON   ☐ Satellite Connection
- ☐ Transmitter Controls OK   ☐ Aircraft Power ON   ☐ Applications / Systems ON

PREFLIGHT NOTES:

_____

SESSION FLIGHT INTERVALS	FLIGHT TIMES		
	START	STOP	TOTAL
1.			
2.			
3.			
4.			
5.			
6.			
7.			
TOTAL HOURS FOR SESSION			
TOTAL FORWARD			
TOTAL TO DATE			

FLIGHT MAP

POSTFLIGHT NOTES / JOURNAL ENTRIES:

I CERTIFY THAT THE FOREGOING ENTRIES ARE TRUE AND CORRECT:

PILOT: _____   SPOTTER: _____

UAS Pilot Log: Unmanned Aircraft Systems Logbook for Drone Pilots and Operators

FLIGHT ID / NO.: _____ Month _____ Day _____ Year _____

FLIGHT LOCATION: _____ Weather: _____

UNMANNED AIRCRAFT SYSTEM: **CREW:**

MANUFACTURER: _____ PILOT: _____

MODEL NUMBER: _____ SPOTTER: _____

PREFLIGHT CHECKLIST:
- ☐ Batteries Charged & Secure
- ☐ Aircraft Hardware OK
- ☐ Equipment & Gear OK
- ☐ Transmitter Controls OK
- ☐ Props OK & Tight
- ☐ Software / Firmware Update
- ☐ Transmitter Control Power ON
- ☐ Aircraft Power ON
- ☐ Compass Calibration
- ☐ Camera / FPV ON
- ☐ Satellite Connection
- ☐ Applications / Systems ON

PREFLIGHT NOTES:
_____

SESSION FLIGHT INTERVALS	FLIGHT TIMES		
	START	STOP	TOTAL
1.			
2.			
3.			
4.			
5.			
6.			
7.			
TOTAL HOURS FOR SESSION			
TOTAL FORWARD			
TOTAL TO DATE			

FLIGHT MAP

POSTFLIGHT NOTES / JOURNAL ENTRIES:

I CERTIFY THAT THE FOREGOING ENTRIES ARE TRUE AND CORRECT:

PILOT: _____ SPOTTER: _____

droneprep.com

UAS Pilot Log: Unmanned Aircraft Systems Logbook for Drone Pilots and Operators

FLIGHT ID / NO.: _____   Month _____ Day _____ Year _____

FLIGHT LOCATION: _____   Weather: _____

UNMANNED AIRCRAFT SYSTEM: **CREW:**

MANUFACTURER: _____   PILOT: _____

MODEL NUMBER: _____   SPOTTER: _____

PREFLIGHT CHECKLIST:

- ☐ Batteries Charged & Secure   ☐ Props OK & Tight   ☐ Compass Calibration
- ☐ Aircraft Hardware OK   ☐ Software / Firmware Update   ☐ Camera / FPV ON
- ☐ Equipment & Gear OK   ☐ Transmitter Control Power ON   ☐ Satellite Connection
- ☐ Transmitter Controls OK   ☐ Aircraft Power ON   ☐ Applications / Systems ON

PREFLIGHT NOTES:

SESSION FLIGHT INTERVALS	FLIGHT TIMES		
	START	STOP	TOTAL
1.			
2.			
3.			
4.			
5.			
6.			
7.			
TOTAL HOURS FOR SESSION			
TOTAL FORWARD			
TOTAL TO DATE			

FLIGHT MAP

POSTFLIGHT NOTES / JOURNAL ENTRIES:

I CERTIFY THAT THE FOREGOING ENTRIES ARE TRUE AND CORRECT:

PILOT: _____   SPOTTER: _____

droneprep.com

UAS Pilot Log: Unmanned Aircraft Systems Logbook for Drone Pilots and Operators

FLIGHT ID / NO.: _____

Month _____ Day _____ Year _____

FLIGHT LOCATION: _____

Weather: _____

UNMANNED AIRCRAFT SYSTEM:

CREW:

MANUFACTURER: _____

PILOT: _____

MODEL NUMBER: _____

SPOTTER: _____

PREFLIGHT CHECKLIST:
- ☐ Batteries Charged & Secure
- ☐ Aircraft Hardware OK
- ☐ Equipment & Gear OK
- ☐ Transmitter Controls OK
- ☐ Props OK & Tight
- ☐ Software / Firmware Update
- ☐ Transmitter Control Power ON
- ☐ Aircraft Power ON
- ☐ Compass Calibration
- ☐ Camera / FPV ON
- ☐ Satellite Connection
- ☐ Applications / Systems ON

PREFLIGHT NOTES:

SESSION FLIGHT INTERVALS	FLIGHT TIMES		
	START	STOP	TOTAL
1.			
2.			
3.			
4.			
5.			
6.			
7.			
TOTAL HOURS FOR SESSION			
TOTAL FORWARD			
TOTAL TO DATE			

FLIGHT MAP

POSTFLIGHT NOTES / JOURNAL ENTRIES:

I CERTIFY THAT THE FOREGOING ENTRIES ARE TRUE AND CORRECT:

PILOT: _____

SPOTTER: _____

droneprep.com

UAS Pilot Log: Unmanned Aircraft Systems Logbook for Drone Pilots and Operators

LIGHT ID / NO.: _____ Month _____ Day _____ Year _____

FLIGHT LOCATION: _____ Weather: _____

UNMANNED AIRCRAFT SYSTEM: **CREW:**

MANUFACTURER: _____ PILOT: _____

MODEL NUMBER: _____ SPOTTER: _____

PREFLIGHT CHECKLIST:

- ☐ Batteries Charged & Secure
- ☐ Aircraft Hardware OK
- ☐ Equipment & Gear OK
- ☐ Transmitter Controls OK

- ☐ Props OK & Tight
- ☐ Software / Firmware Update
- ☐ Transmitter Control Power ON
- ☐ Aircraft Power ON

- ☐ Compass Calibration
- ☐ Camera / FPV ON
- ☐ Satellite Connection
- ☐ Applications / Systems ON

PREFLIGHT NOTES:

SESSION FLIGHT INTERVALS	FLIGHT TIMES		
	START	STOP	TOTAL
1.			
2.			
3.			
4.			
5.			
6.			
7.			
TOTAL HOURS FOR SESSION			
TOTAL FORWARD			
TOTAL TO DATE			

FLIGHT MAP

POSTFLIGHT NOTES / JOURNAL ENTRIES:

I CERTIFY THAT THE FOREGOING ENTRIES ARE TRUE AND CORRECT:

PILOT: _____ SPOTTER: _____

UAS Pilot Log: Unmanned Aircraft Systems Logbook for Drone Pilots and Operators

FLIGHT ID / NO.: _____

Month _____ Day _____ Year _____

FLIGHT LOCATION: _____

Weather: _____

UNMANNED AIRCRAFT SYSTEM:

CREW:

MANUFACTURER: _____

PILOT: _____

MODEL NUMBER: _____

SPOTTER: _____

PREFLIGHT CHECKLIST:
- ☐ Batteries Charged & Secure
- ☐ Aircraft Hardware OK
- ☐ Equipment & Gear OK
- ☐ Transmitter Controls OK

- ☐ Props OK & Tight
- ☐ Software / Firmware Update
- ☐ Transmitter Control Power ON
- ☐ Aircraft Power ON

- ☐ Compass Calibration
- ☐ Camera / FPV ON
- ☐ Satellite Connection
- ☐ Applications / Systems ON

PREFLIGHT NOTES:

SESSION FLIGHT INTERVALS	FLIGHT TIMES		
	START	STOP	TOTAL
1.			
2.			
3.			
4.			
5.			
6.			
7.			
TOTAL HOURS FOR SESSION			
TOTAL FORWARD			
TOTAL TO DATE			

FLIGHT MAP

POSTFLIGHT NOTES / JOURNAL ENTRIES:

I CERTIFY THAT THE FOREGOING ENTRIES ARE TRUE AND CORRECT:

PILOT: _____

SPOTTER: _____

UAS Pilot Log: Unmanned Aircraft Systems Logbook for Drone Pilots and Operators

FLIGHT ID / NO.: _____

Month _____ Day _____ Year _____

FLIGHT LOCATION: _____

Weather: _____

UNMANNED AIRCRAFT SYSTEM:

CREW:

MANUFACTURER: _____

PILOT: _____

MODEL NUMBER: _____

SPOTTER: _____

PREFLIGHT CHECKLIST:

- ☐ Batteries Charged & Secure
- ☐ Aircraft Hardware OK
- ☐ Equipment & Gear OK
- ☐ Transmitter Controls OK
- ☐ Props OK & Tight
- ☐ Software / Firmware Update
- ☐ Transmitter Control Power ON
- ☐ Aircraft Power ON
- ☐ Compass Calibration
- ☐ Camera / FPV ON
- ☐ Satellite Connection
- ☐ Applications / Systems ON

PREFLIGHT NOTES:

SESSION FLIGHT INTERVALS	FLIGHT TIMES		
	START	STOP	TOTAL
1.			
2.			
3.			
4.			
5.			
6.			
7.			
TOTAL HOURS FOR SESSION			
TOTAL FORWARD			
TOTAL TO DATE			

FLIGHT MAP

POSTFLIGHT NOTES / JOURNAL ENTRIES:

I CERTIFY THAT THE FOREGOING ENTRIES ARE TRUE AND CORRECT:

PILOT: _____

SPOTTER: _____

droneprep.com

UAS Pilot Log: Unmanned Aircraft Systems Logbook for Drone Pilots and Operators

FLIGHT ID / NO.: _____

Month _____ Day _____ Year _____

FLIGHT LOCATION: _____

Weather: _____

UNMANNED AIRCRAFT SYSTEM:

CREW:

MANUFACTURER: _____

PILOT: _____

MODEL NUMBER: _____

SPOTTER: _____

PREFLIGHT CHECKLIST:

☐ Batteries Charged & Secure
☐ Aircraft Hardware OK
☐ Equipment & Gear OK
☐ Transmitter Controls OK

☐ Props OK & Tight
☐ Software / Firmware Update
☐ Transmitter Control Power ON
☐ Aircraft Power ON

☐ Compass Calibration
☐ Camera / FPV ON
☐ Satellite Connection
☐ Applications / Systems ON

PREFLIGHT NOTES:

SESSION FLIGHT INTERVALS	FLIGHT TIMES		
	START	STOP	TOTAL
1.			
2.			
3.			
4.			
5.			
6.			
7.			
TOTAL HOURS FOR SESSION			
TOTAL FORWARD			
TOTAL TO DATE			

FLIGHT MAP

POSTFLIGHT NOTES / JOURNAL ENTRIES:

I CERTIFY THAT THE FOREGOING ENTRIES ARE TRUE AND CORRECT:

PILOT: _____   SPOTTER: _____

UAS Pilot Log: Unmanned Aircraft Systems Logbook for Drone Pilots and Operators

FLIGHT ID / NO.: _____   Month _____ Day _____ Year _____

FLIGHT LOCATION: _____   Weather: _____

UNMANNED AIRCRAFT SYSTEM: **CREW:**

MANUFACTURER: _____   PILOT: _____

MODEL NUMBER: _____   SPOTTER: _____

PREFLIGHT CHECKLIST:

- ☐ Batteries Charged & Secure
- ☐ Aircraft Hardware OK
- ☐ Equipment & Gear OK
- ☐ Transmitter Controls OK
- ☐ Props OK & Tight
- ☐ Software / Firmware Update
- ☐ Transmitter Control Power ON
- ☐ Aircraft Power ON
- ☐ Compass Calibration
- ☐ Camera / FPV ON
- ☐ Satellite Connection
- ☐ Applications / Systems ON

PREFLIGHT NOTES:

_____

SESSION FLIGHT INTERVALS	FLIGHT TIMES		
	START	STOP	TOTAL
1.			
2.			
3.			
4.			
5.			
6.			
7.			
TOTAL HOURS FOR SESSION			
TOTAL FORWARD			
TOTAL TO DATE			

FLIGHT MAP

POSTFLIGHT NOTES / JOURNAL ENTRIES:

I CERTIFY THAT THE FOREGOING ENTRIES ARE TRUE AND CORRECT:

PILOT: _____   SPOTTER: _____

droneprep.com

UAS Pilot Log: Unmanned Aircraft Systems Logbook for Drone Pilots and Operators

FLIGHT ID / NO.: _____

Month _____ Day _____ Year _____

FLIGHT LOCATION: _____

Weather: _____

UNMANNED AIRCRAFT SYSTEM:

CREW:

MANUFACTURER: _____

PILOT: _____

MODEL NUMBER: _____

SPOTTER: _____

PREFLIGHT CHECKLIST:

- ☐ Batteries Charged & Secure
- ☐ Aircraft Hardware OK
- ☐ Equipment & Gear OK
- ☐ Transmitter Controls OK

- ☐ Props OK & Tight
- ☐ Software / Firmware Update
- ☐ Transmitter Control Power ON
- ☐ Aircraft Power ON

- ☐ Compass Calibration
- ☐ Camera / FPV ON
- ☐ Satellite Connection
- ☐ Applications / Systems ON

PREFLIGHT NOTES:

SESSION FLIGHT INTERVALS	FLIGHT TIMES		
	START	STOP	TOTAL
1.			
2.			
3.			
4.			
5.			
6.			
7.			
TOTAL HOURS FOR SESSION			
TOTAL FORWARD			
TOTAL TO DATE			

FLIGHT MAP

POSTFLIGHT NOTES / JOURNAL ENTRIES:

I CERTIFY THAT THE FOREGOING ENTRIES ARE TRUE AND CORRECT:

PILOT: _____   SPOTTER: _____

UAS Pilot Log: Unmanned Aircraft Systems Logbook for Drone Pilots and Operators

FLIGHT ID / NO.: _____

Month _____ Day _____ Year _____

FLIGHT LOCATION: _____

Weather: _____

UNMANNED AIRCRAFT SYSTEM:

CREW:

MANUFACTURER: _____

PILOT: _____

MODEL NUMBER: _____

SPOTTER: _____

PREFLIGHT CHECKLIST:

- ☐ Batteries Charged & Secure
- ☐ Aircraft Hardware OK
- ☐ Equipment & Gear OK
- ☐ Transmitter Controls OK
- ☐ Props OK & Tight
- ☐ Software / Firmware Update
- ☐ Transmitter Control Power ON
- ☐ Aircraft Power ON
- ☐ Compass Calibration
- ☐ Camera / FPV ON
- ☐ Satellite Connection
- ☐ Applications / Systems ON

PREFLIGHT NOTES:

SESSION FLIGHT INTERVALS	FLIGHT TIMES		
	START	STOP	TOTAL
1.			
2.			
3.			
4.			
5.			
6.			
7.			
TOTAL HOURS FOR SESSION			
TOTAL FORWARD			
TOTAL TO DATE			

FLIGHT MAP

POSTFLIGHT NOTES / JOURNAL ENTRIES:

I CERTIFY THAT THE FOREGOING ENTRIES ARE TRUE AND CORRECT:

PILOT: _____   SPOTTER: _____

droneprep.com

UAS Pilot Log: Unmanned Aircraft Systems Logbook for Drone Pilots and Operators

FLIGHT ID / NO.: _____ Month _____ Day _____ Year _____

FLIGHT LOCATION: _____ Weather: _____

UNMANNED AIRCRAFT SYSTEM: **CREW:**

MANUFACTURER: _____   PILOT: _____

MODEL NUMBER: _____   SPOTTER: _____

PREFLIGHT CHECKLIST:

- ☐ Batteries Charged & Secure
- ☐ Aircraft Hardware OK
- ☐ Equipment & Gear OK
- ☐ Transmitter Controls OK
- ☐ Props OK & Tight
- ☐ Software / Firmware Update
- ☐ Transmitter Control Power ON
- ☐ Aircraft Power ON
- ☐ Compass Calibration
- ☐ Camera / FPV ON
- ☐ Satellite Connection
- ☐ Applications / Systems ON

PREFLIGHT NOTES:

_____

SESSION FLIGHT INTERVALS	FLIGHT TIMES		
	START	STOP	TOTAL
1.			
2.			
3.			
4.			
5.			
6.			
7.			
TOTAL HOURS FOR SESSION			
TOTAL FORWARD			
TOTAL TO DATE			

FLIGHT MAP

POSTFLIGHT NOTES / JOURNAL ENTRIES:

I CERTIFY THAT THE FOREGOING ENTRIES ARE TRUE AND CORRECT:

PILOT: _____   SPOTTER: _____

UAS Pilot Log: Unmanned Aircraft Systems Logbook for Drone Pilots and Operators

FLIGHT ID / NO.: _____

Month _____ Day _____ Year _____

FLIGHT LOCATION: _____

Weather: _____

UNMANNED AIRCRAFT SYSTEM:

MANUFACTURER: _____

MODEL NUMBER: _____

CREW:

PILOT: _____

SPOTTER: _____

PREFLIGHT CHECKLIST:

- ☐ Batteries Charged & Secure
- ☐ Aircraft Hardware OK
- ☐ Equipment & Gear OK
- ☐ Transmitter Controls OK
- ☐ Props OK & Tight
- ☐ Software / Firmware Update
- ☐ Transmitter Control Power ON
- ☐ Aircraft Power ON
- ☐ Compass Calibration
- ☐ Camera / FPV ON
- ☐ Satellite Connection
- ☐ Applications / Systems ON

PREFLIGHT NOTES:

SESSION FLIGHT INTERVALS	FLIGHT TIMES		
	START	STOP	TOTAL
1.			
2.			
3.			
4.			
5.			
6.			
7.			
TOTAL HOURS FOR SESSION			
TOTAL FORWARD			
TOTAL TO DATE			

FLIGHT MAP

POSTFLIGHT NOTES / JOURNAL ENTRIES:

I CERTIFY THAT THE FOREGOING ENTRIES ARE TRUE AND CORRECT:

PILOT: _____ SPOTTER: _____

UAS Pilot Log: Unmanned Aircraft Systems Logbook for Drone Pilots and Operators

FLIGHT ID / NO.: _____

Month _____ Day _____ Year _____

FLIGHT LOCATION: _____

Weather: _____

UNMANNED AIRCRAFT SYSTEM:

CREW:

MANUFACTURER: _____

PILOT: _____

MODEL NUMBER: _____

SPOTTER: _____

PREFLIGHT CHECKLIST:
- ☐ Batteries Charged & Secure
- ☐ Aircraft Hardware OK
- ☐ Equipment & Gear OK
- ☐ Transmitter Controls OK
- ☐ Props OK & Tight
- ☐ Software / Firmware Update
- ☐ Transmitter Control Power ON
- ☐ Aircraft Power ON
- ☐ Compass Calibration
- ☐ Camera / FPV ON
- ☐ Satellite Connection
- ☐ Applications / Systems ON

PREFLIGHT NOTES:

SESSION FLIGHT INTERVALS	FLIGHT TIMES		
	START	STOP	TOTAL
1.			
2.			
3.			
4.			
5.			
6.			
7.			
TOTAL HOURS FOR SESSION			
TOTAL FORWARD			
TOTAL TO DATE			

FLIGHT MAP

POSTFLIGHT NOTES / JOURNAL ENTRIES:

I CERTIFY THAT THE FOREGOING ENTRIES ARE TRUE AND CORRECT:

PILOT: _____   SPOTTER: _____

UAS Pilot Log: Unmanned Aircraft Systems Logbook for Drone Pilots and Operators

FLIGHT ID / NO.: _____   Month _____ Day _____ Year _____

FLIGHT LOCATION: _____   Weather: _____

UNMANNED AIRCRAFT SYSTEM: CREW:

MANUFACTURER: _____   PILOT: _____

MODEL NUMBER: _____   SPOTTER: _____

PREFLIGHT CHECKLIST:

- ☐ Batteries Charged & Secure   ☐ Props OK & Tight   ☐ Compass Calibration
- ☐ Aircraft Hardware OK   ☐ Software / Firmware Update   ☐ Camera / FPV ON
- ☐ Equipment & Gear OK   ☐ Transmitter Control Power ON ☐ Satellite Connection
- ☐ Transmitter Controls OK   ☐ Aircraft Power ON   ☐ Applications / Systems ON

PREFLIGHT NOTES:

SESSION FLIGHT INTERVALS	FLIGHT TIMES		
	START	STOP	TOTAL
1.			
2.			
3.			
4.			
5.			
6.			
7.			
TOTAL HOURS FOR SESSION			
TOTAL FORWARD			
TOTAL TO DATE			

FLIGHT MAP

POSTFLIGHT NOTES / JOURNAL ENTRIES:

I CERTIFY THAT THE FOREGOING ENTRIES ARE TRUE AND CORRECT:

PILOT: _____   SPOTTER: _____

droneprep.com

UAS Pilot Log: Unmanned Aircraft Systems Logbook for Drone Pilots and Operators

FLIGHT ID / NO.: _____

Month _____ Day _____ Year _____

FLIGHT LOCATION: _____

Weather: _____

UNMANNED AIRCRAFT SYSTEM:

CREW:

MANUFACTURER: _____

PILOT: _____

MODEL NUMBER: _____

SPOTTER: _____

PREFLIGHT CHECKLIST:

- ☐ Batteries Charged & Secure
- ☐ Aircraft Hardware OK
- ☐ Equipment & Gear OK
- ☐ Transmitter Controls OK

- ☐ Props OK & Tight
- ☐ Software / Firmware Update
- ☐ Transmitter Control Power ON
- ☐ Aircraft Power ON

- ☐ Compass Calibration
- ☐ Camera / FPV ON
- ☐ Satellite Connection
- ☐ Applications / Systems ON

PREFLIGHT NOTES:

SESSION FLIGHT INTERVALS	FLIGHT TIMES		
	START	STOP	TOTAL
1.			
2.			
3.			
4.			
5.			
6.			
7.			
TOTAL HOURS FOR SESSION			
TOTAL FORWARD			
TOTAL TO DATE			

FLIGHT MAP

POSTFLIGHT NOTES / JOURNAL ENTRIES:

I CERTIFY THAT THE FOREGOING ENTRIES ARE TRUE AND CORRECT:

PILOT: _____

SPOTTER: _____

UAS Pilot Log: Unmanned Aircraft Systems Logbook for Drone Pilots and Operators

FLIGHT ID / NO.: _____

Month _____ Day _____ Year _____

FLIGHT LOCATION: _____

Weather: _____

UNMANNED AIRCRAFT SYSTEM:

MANUFACTURER: _____

MODEL NUMBER: _____

CREW:

PILOT: _____

SPOTTER: _____

PREFLIGHT CHECKLIST:

- ☐ Batteries Charged & Secure
- ☐ Aircraft Hardware OK
- ☐ Equipment & Gear OK
- ☐ Transmitter Controls OK
- ☐ Props OK & Tight
- ☐ Software / Firmware Update
- ☐ Transmitter Control Power ON
- ☐ Aircraft Power ON
- ☐ Compass Calibration
- ☐ Camera / FPV ON
- ☐ Satellite Connection
- ☐ Applications / Systems ON

PREFLIGHT NOTES:

SESSION FLIGHT INTERVALS	FLIGHT TIMES		
	START	STOP	TOTAL
1.			
2.			
3.			
4.			
5.			
6.			
7.			
TOTAL HOURS FOR SESSION			
TOTAL FORWARD			
TOTAL TO DATE			

FLIGHT MAP

POSTFLIGHT NOTES / JOURNAL ENTRIES:

I CERTIFY THAT THE FOREGOING ENTRIES ARE TRUE AND CORRECT:

PILOT: _____   SPOTTER: _____

droneprep.com

UAS Pilot Log: Unmanned Aircraft Systems Logbook for Drone Pilots and Operators

FLIGHT ID / NO.: _____

Month _____ Day _____ Year _____

FLIGHT LOCATION: _____

Weather: _____

UNMANNED AIRCRAFT SYSTEM:

CREW:

MANUFACTURER: _____

PILOT: _____

MODEL NUMBER: _____

SPOTTER: _____

PREFLIGHT CHECKLIST:

- ☐ Batteries Charged & Secure
- ☐ Aircraft Hardware OK
- ☐ Equipment & Gear OK
- ☐ Transmitter Controls OK

- ☐ Props OK & Tight
- ☐ Software / Firmware Update
- ☐ Transmitter Control Power ON
- ☐ Aircraft Power ON

- ☐ Compass Calibration
- ☐ Camera / FPV ON
- ☐ Satellite Connection
- ☐ Applications / Systems ON

PREFLIGHT NOTES:

SESSION FLIGHT INTERVALS	FLIGHT TIMES		
	START	STOP	TOTAL
1.			
2.			
3.			
4.			
5.			
6.			
7.			
TOTAL HOURS FOR SESSION			
TOTAL FORWARD			
TOTAL TO DATE			

FLIGHT MAP

POSTFLIGHT NOTES / JOURNAL ENTRIES:

I CERTIFY THAT THE FOREGOING ENTRIES ARE TRUE AND CORRECT:

PILOT: _____ SPOTTER: _____

UAS Pilot Log: Unmanned Aircraft Systems Logbook for Drone Pilots and Operators

FLIGHT ID / NO.: _____   Month _____ Day _____ Year _____

FLIGHT LOCATION: _____   Weather: _____

UNMANNED AIRCRAFT SYSTEM: **CREW:**

MANUFACTURER: _____   PILOT: _____

MODEL NUMBER: _____   SPOTTER: _____

PREFLIGHT CHECKLIST:

- ☐ Batteries Charged & Secure
- ☐ Aircraft Hardware OK
- ☐ Equipment & Gear OK
- ☐ Transmitter Controls OK
- ☐ Props OK & Tight
- ☐ Software / Firmware Update
- ☐ Transmitter Control Power ON
- ☐ Aircraft Power ON
- ☐ Compass Calibration
- ☐ Camera / FPV ON
- ☐ Satellite Connection
- ☐ Applications / Systems ON

PREFLIGHT NOTES:

SESSION FLIGHT INTERVALS	FLIGHT TIMES		
	START	STOP	TOTAL
1.			
2.			
3.			
4.			
5.			
6.			
7.			
TOTAL HOURS FOR SESSION			
TOTAL FORWARD			
TOTAL TO DATE			

FLIGHT MAP

POSTFLIGHT NOTES / JOURNAL ENTRIES:

I CERTIFY THAT THE FOREGOING ENTRIES ARE TRUE AND CORRECT:

PILOT: _____   SPOTTER: _____

UAS Pilot Log: Unmanned Aircraft Systems Logbook for Drone Pilots and Operators

FLIGHT ID / NO.: _____   Month _____ Day _____ Year _____

FLIGHT LOCATION: _____   Weather: _____

UNMANNED AIRCRAFT SYSTEM: **CREW:**

MANUFACTURER: _____   PILOT: _____

MODEL NUMBER: _____   SPOTTER: _____

PREFLIGHT CHECKLIST:
- ☐ Batteries Charged & Secure
- ☐ Aircraft Hardware OK
- ☐ Equipment & Gear OK
- ☐ Transmitter Controls OK
- ☐ Props OK & Tight
- ☐ Software / Firmware Update
- ☐ Transmitter Control Power ON
- ☐ Aircraft Power ON
- ☐ Compass Calibration
- ☐ Camera / FPV ON
- ☐ Satellite Connection
- ☐ Applications / Systems ON

PREFLIGHT NOTES:

SESSION FLIGHT INTERVALS	FLIGHT TIMES		
	START	STOP	TOTAL
1.			
2.			
3.			
4.			
5.			
6.			
7.			
TOTAL HOURS FOR SESSION			
TOTAL FORWARD			
TOTAL TO DATE			

FLIGHT MAP

POSTFLIGHT NOTES / JOURNAL ENTRIES:

I CERTIFY THAT THE FOREGOING ENTRIES ARE TRUE AND CORRECT:

PILOT: _____   SPOTTER: _____

UAS Pilot Log: Unmanned Aircraft Systems Logbook for Drone Pilots and Operators

FLIGHT ID / NO.: _____

Month _____ Day _____ Year _____

FLIGHT LOCATION: _____

Weather: _____

UNMANNED AIRCRAFT SYSTEM:

CREW:

MANUFACTURER: _____

PILOT: _____

MODEL NUMBER: _____

SPOTTER: _____

PREFLIGHT CHECKLIST:

- ☐ Batteries Charged & Secure
- ☐ Aircraft Hardware OK
- ☐ Equipment & Gear OK
- ☐ Transmitter Controls OK

- ☐ Props OK & Tight
- ☐ Software / Firmware Update
- ☐ Transmitter Control Power ON
- ☐ Aircraft Power ON

- ☐ Compass Calibration
- ☐ Camera / FPV ON
- ☐ Satellite Connection
- ☐ Applications / Systems ON

PREFLIGHT NOTES:

SESSION FLIGHT INTERVALS	FLIGHT TIMES		
	START	STOP	TOTAL
1.			
2.			
3.			
4.			
5.			
6.			
7.			
TOTAL HOURS FOR SESSION			
TOTAL FORWARD			
TOTAL TO DATE			

FLIGHT MAP

POSTFLIGHT NOTES / JOURNAL ENTRIES:

I CERTIFY THAT THE FOREGOING ENTRIES ARE TRUE AND CORRECT:

PILOT: _____   SPOTTER: _____

UAS Pilot Log: Unmanned Aircraft Systems Logbook for Drone Pilots and Operators

FLIGHT ID / NO.: _____

FLIGHT LOCATION: _____

Month _____ Day _____ Year _____

Weather: _____

UNMANNED AIRCRAFT SYSTEM:

MANUFACTURER: _____

MODEL NUMBER: _____

CREW:

PILOT: _____

SPOTTER: _____

PREFLIGHT CHECKLIST:

- ☐ Batteries Charged & Secure
- ☐ Aircraft Hardware OK
- ☐ Equipment & Gear OK
- ☐ Transmitter Controls OK

- ☐ Props OK & Tight
- ☐ Software / Firmware Update
- ☐ Transmitter Control Power ON
- ☐ Aircraft Power ON

- ☐ Compass Calibration
- ☐ Camera / FPV ON
- ☐ Satellite Connection
- ☐ Applications / Systems ON

PREFLIGHT NOTES:

SESSION FLIGHT INTERVALS	FLIGHT TIMES		
	START	STOP	TOTAL
1.			
2.			
3.			
4.			
5.			
6.			
7.			
TOTAL HOURS FOR SESSION			
TOTAL FORWARD			
TOTAL TO DATE			

FLIGHT MAP

POSTFLIGHT NOTES / JOURNAL ENTRIES:

I CERTIFY THAT THE FOREGOING ENTRIES ARE TRUE AND CORRECT:

PILOT: _____   SPOTTER: _____

UAS Pilot Log: Unmanned Aircraft Systems Logbook for Drone Pilots and Operators

FLIGHT ID / NO.: _____   Month _____ Day _____ Year _____

FLIGHT LOCATION: _____   Weather: _____

UNMANNED AIRCRAFT SYSTEM: **CREW:**

MANUFACTURER: _____   PILOT: _____

MODEL NUMBER: _____   SPOTTER: _____

PREFLIGHT CHECKLIST:

- ☐ Batteries Charged & Secure   ☐ Props OK & Tight   ☐ Compass Calibration
- ☐ Aircraft Hardware OK   ☐ Software / Firmware Update   ☐ Camera / FPV ON
- ☐ Equipment & Gear OK   ☐ Transmitter Control Power ON   ☐ Satellite Connection
- ☐ Transmitter Controls OK   ☐ Aircraft Power ON   ☐ Applications / Systems ON

PREFLIGHT NOTES:

SESSION FLIGHT INTERVALS	FLIGHT TIMES		
	START	STOP	TOTAL
1.			
2.			
3.			
4.			
5.			
6.			
7.			
TOTAL HOURS FOR SESSION			
TOTAL FORWARD			
TOTAL TO DATE			

FLIGHT MAP

POSTFLIGHT NOTES / JOURNAL ENTRIES:

I CERTIFY THAT THE FOREGOING ENTRIES ARE TRUE AND CORRECT:

PILOT: _____   SPOTTER: _____

UAS Pilot Log: Unmanned Aircraft Systems Logbook for Drone Pilots and Operators

FLIGHT ID / NO.: _____

Month _____ Day _____ Year _____

FLIGHT LOCATION: _____

Weather: _____

UNMANNED AIRCRAFT SYSTEM:

CREW:

MANUFACTURER: _____

PILOT: _____

MODEL NUMBER: _____

SPOTTER: _____

PREFLIGHT CHECKLIST:

- ☐ Batteries Charged & Secure
- ☐ Aircraft Hardware OK
- ☐ Equipment & Gear OK
- ☐ Transmitter Controls OK
- ☐ Props OK & Tight
- ☐ Software / Firmware Update
- ☐ Transmitter Control Power ON
- ☐ Aircraft Power ON
- ☐ Compass Calibration
- ☐ Camera / FPV ON
- ☐ Satellite Connection
- ☐ Applications / Systems ON

PREFLIGHT NOTES:

SESSION FLIGHT INTERVALS	FLIGHT TIMES		
	START	STOP	TOTAL
1.			
2.			
3.			
4.			
5.			
6.			
7.			
TOTAL HOURS FOR SESSION			
TOTAL FORWARD			
TOTAL TO DATE			

FLIGHT MAP

POSTFLIGHT NOTES / JOURNAL ENTRIES:

I CERTIFY THAT THE FOREGOING ENTRIES ARE TRUE AND CORRECT:

PILOT: _____ SPOTTER: _____

UAS Pilot Log: Unmanned Aircraft Systems Logbook for Drone Pilots and Operators

FLIGHT ID / NO.: _____

Month _____ Day _____ Year _____

FLIGHT LOCATION: _____

Weather: _____

UNMANNED AIRCRAFT SYSTEM:

CREW:

MANUFACTURER: _____

PILOT: _____

MODEL NUMBER: _____

SPOTTER: _____

PREFLIGHT CHECKLIST:

- ☐ Batteries Charged & Secure
- ☐ Aircraft Hardware OK
- ☐ Equipment & Gear OK
- ☐ Transmitter Controls OK
- ☐ Props OK & Tight
- ☐ Software / Firmware Update
- ☐ Transmitter Control Power ON
- ☐ Aircraft Power ON
- ☐ Compass Calibration
- ☐ Camera / FPV ON
- ☐ Satellite Connection
- ☐ Applications / Systems ON

PREFLIGHT NOTES:

SESSION FLIGHT INTERVALS	FLIGHT TIMES		
	START	STOP	TOTAL
1.			
2.			
3.			
4.			
5.			
6.			
7.			
TOTAL HOURS FOR SESSION			
TOTAL FORWARD			
TOTAL TO DATE			

FLIGHT MAP

POSTFLIGHT NOTES / JOURNAL ENTRIES:

I CERTIFY THAT THE FOREGOING ENTRIES ARE TRUE AND CORRECT:

PILOT: _____

SPOTTER: _____

droneprep.com

UAS Pilot Log: Unmanned Aircraft Systems Logbook for Drone Pilots and Operators

FLIGHT ID / NO.: _____   Month _____ Day _____ Year _____

FLIGHT LOCATION: _____   Weather: _____

UNMANNED AIRCRAFT SYSTEM: **CREW:**

MANUFACTURER: _____   PILOT: _____

MODEL NUMBER: _____   SPOTTER: _____

PREFLIGHT CHECKLIST:
- ☐ Batteries Charged & Secure
- ☐ Aircraft Hardware OK
- ☐ Equipment & Gear OK
- ☐ Transmitter Controls OK
- ☐ Props OK & Tight
- ☐ Software / Firmware Update
- ☐ Transmitter Control Power ON
- ☐ Aircraft Power ON
- ☐ Compass Calibration
- ☐ Camera / FPV ON
- ☐ Satellite Connection
- ☐ Applications / Systems ON

PREFLIGHT NOTES:

SESSION FLIGHT INTERVALS	FLIGHT TIMES		
	START	STOP	TOTAL
1.			
2.			
3.			
4.			
5.			
6.			
7.			
TOTAL HOURS FOR SESSION			
TOTAL FORWARD			
TOTAL TO DATE			

FLIGHT MAP

POSTFLIGHT NOTES / JOURNAL ENTRIES:

I CERTIFY THAT THE FOREGOING ENTRIES ARE TRUE AND CORRECT:

PILOT: _____   SPOTTER: _____

UAS Pilot Log: Unmanned Aircraft Systems Logbook for Drone Pilots and Operators

FLIGHT ID / NO.: _____   Month _____ Day _____ Year _____

FLIGHT LOCATION: _____   Weather: _____

UNMANNED AIRCRAFT SYSTEM: CREW:

MANUFACTURER: _____   PILOT: _____

MODEL NUMBER: _____   SPOTTER: _____

PREFLIGHT CHECKLIST:

- ☐ Batteries Charged & Secure
- ☐ Aircraft Hardware OK
- ☐ Equipment & Gear OK
- ☐ Transmitter Controls OK

- ☐ Props OK & Tight
- ☐ Software / Firmware Update
- ☐ Transmitter Control Power ON
- ☐ Aircraft Power ON

- ☐ Compass Calibration
- ☐ Camera / FPV ON
- ☐ Satellite Connection
- ☐ Applications / Systems ON

PREFLIGHT NOTES:

SESSION FLIGHT INTERVALS	FLIGHT TIMES		
	START	STOP	TOTAL
1.			
2.			
3.			
4.			
5.			
6.			
7.			
TOTAL HOURS FOR SESSION			
TOTAL FORWARD			
TOTAL TO DATE			

FLIGHT MAP

POSTFLIGHT NOTES / JOURNAL ENTRIES:

I CERTIFY THAT THE FOREGOING ENTRIES ARE TRUE AND CORRECT:

PILOT: _____   SPOTTER: _____

droneprep.com

UAS Pilot Log: Unmanned Aircraft Systems Logbook for Drone Pilots and Operators

FLIGHT ID / NO.: _____   Month _____ Day _____ Year _____

FLIGHT LOCATION: _____   Weather: _____

UNMANNED AIRCRAFT SYSTEM: **CREW:**

MANUFACTURER: _____   PILOT: _____

MODEL NUMBER: _____   SPOTTER: _____

PREFLIGHT CHECKLIST:
- ☐ Batteries Charged & Secure   ☐ Props OK & Tight   ☐ Compass Calibration
- ☐ Aircraft Hardware OK   ☐ Software / Firmware Update   ☐ Camera / FPV ON
- ☐ Equipment & Gear OK   ☐ Transmitter Control Power ON   ☐ Satellite Connection
- ☐ Transmitter Controls OK   ☐ Aircraft Power ON   ☐ Applications / Systems ON

PREFLIGHT NOTES:

SESSION FLIGHT INTERVALS	FLIGHT TIMES		
	START	STOP	TOTAL
1.			
2.			
3.			
4.			
5.			
6.			
7.			
TOTAL HOURS FOR SESSION			
TOTAL FORWARD			
TOTAL TO DATE			

FLIGHT MAP

POSTFLIGHT NOTES / JOURNAL ENTRIES:

I CERTIFY THAT THE FOREGOING ENTRIES ARE TRUE AND CORRECT:

PILOT: _____   SPOTTER: _____

UAS Pilot Log: Unmanned Aircraft Systems Logbook for Drone Pilots and Operators

FLIGHT ID / NO.: _____

Month _____ Day _____ Year _____

FLIGHT LOCATION: _____

Weather: _____

UNMANNED AIRCRAFT SYSTEM:

MANUFACTURER: _____

MODEL NUMBER: _____

CREW:

PILOT: _____

SPOTTER: _____

PREFLIGHT CHECKLIST:

- ☐ Batteries Charged & Secure
- ☐ Aircraft Hardware OK
- ☐ Equipment & Gear OK
- ☐ Transmitter Controls OK
- ☐ Props OK & Tight
- ☐ Software / Firmware Update
- ☐ Transmitter Control Power ON
- ☐ Aircraft Power ON
- ☐ Compass Calibration
- ☐ Camera / FPV ON
- ☐ Satellite Connection
- ☐ Applications / Systems ON

PREFLIGHT NOTES:

SESSION FLIGHT INTERVALS	FLIGHT TIMES		
	START	STOP	TOTAL
1.			
2.			
3.			
4.			
5.			
6.			
7.			
TOTAL HOURS FOR SESSION			
TOTAL FORWARD			
TOTAL TO DATE			

FLIGHT MAP

POSTFLIGHT NOTES / JOURNAL ENTRIES:

I CERTIFY THAT THE FOREGOING ENTRIES ARE TRUE AND CORRECT:

PILOT: _____   SPOTTER: _____

droneprep.com

UAS Pilot Log: Unmanned Aircraft Systems Logbook for Drone Pilots and Operators

FLIGHT ID / NO.: _____ Month _____ Day _____ Year _____

FLIGHT LOCATION: _____ Weather: _____

UNMANNED AIRCRAFT SYSTEM: **CREW:**

MANUFACTURER: _____ PILOT: _____

MODEL NUMBER: _____ SPOTTER: _____

PREFLIGHT CHECKLIST:
- ☐ Batteries Charged & Secure
- ☐ Aircraft Hardware OK
- ☐ Equipment & Gear OK
- ☐ Transmitter Controls OK
- ☐ Props OK & Tight
- ☐ Software / Firmware Update
- ☐ Transmitter Control Power ON
- ☐ Aircraft Power ON
- ☐ Compass Calibration
- ☐ Camera / FPV ON
- ☐ Satellite Connection
- ☐ Applications / Systems ON

PREFLIGHT NOTES:

_____

SESSION FLIGHT INTERVALS	FLIGHT TIMES		
	START	STOP	TOTAL
1.			
2.			
3.			
4.			
5.			
6.			
7.			
TOTAL HOURS FOR SESSION			
TOTAL FORWARD			
TOTAL TO DATE			

FLIGHT MAP

POSTFLIGHT NOTES / JOURNAL ENTRIES:

I CERTIFY THAT THE FOREGOING ENTRIES ARE TRUE AND CORRECT:

PILOT: _____ SPOTTER: _____

UAS Pilot Log: Unmanned Aircraft Systems Logbook for Drone Pilots and Operators

FLIGHT ID / NO.: _____   Month _____ Day _____ Year _____

FLIGHT LOCATION: _____   Weather: _____

UNMANNED AIRCRAFT SYSTEM: **CREW:**

MANUFACTURER: _____   PILOT: _____

MODEL NUMBER: _____   SPOTTER: _____

PREFLIGHT CHECKLIST:

- ☐ Batteries Charged & Secure
- ☐ Aircraft Hardware OK
- ☐ Equipment & Gear OK
- ☐ Transmitter Controls OK
- ☐ Props OK & Tight
- ☐ Software / Firmware Update
- ☐ Transmitter Control Power ON
- ☐ Aircraft Power ON
- ☐ Compass Calibration
- ☐ Camera / FPV ON
- ☐ Satellite Connection
- ☐ Applications / Systems ON

PREFLIGHT NOTES:

SESSION FLIGHT INTERVALS	FLIGHT TIMES		
	START	STOP	TOTAL
1.			
2.			
3.			
4.			
5.			
6.			
7.			
TOTAL HOURS FOR SESSION			
TOTAL FORWARD			
TOTAL TO DATE			

FLIGHT MAP

POSTFLIGHT NOTES / JOURNAL ENTRIES:

I CERTIFY THAT THE FOREGOING ENTRIES ARE TRUE AND CORRECT:

PILOT: _____   SPOTTER: _____

droneprep.com

UAS Pilot Log: Unmanned Aircraft Systems Logbook for Drone Pilots and Operators

FLIGHT ID / NO.: _____

Month _____ Day _____ Year _____

FLIGHT LOCATION: _____

Weather: _____

UNMANNED AIRCRAFT SYSTEM:

CREW:

MANUFACTURER: _____

PILOT: _____

MODEL NUMBER: _____

SPOTTER: _____

PREFLIGHT CHECKLIST:

- ☐ Batteries Charged & Secure
- ☐ Aircraft Hardware OK
- ☐ Equipment & Gear OK
- ☐ Transmitter Controls OK
- ☐ Props OK & Tight
- ☐ Software / Firmware Update
- ☐ Transmitter Control Power ON
- ☐ Aircraft Power ON
- ☐ Compass Calibration
- ☐ Camera / FPV ON
- ☐ Satellite Connection
- ☐ Applications / Systems ON

PREFLIGHT NOTES:

SESSION FLIGHT INTERVALS	FLIGHT TIMES		
	START	STOP	TOTAL
1.			
2.			
3.			
4.			
5.			
6.			
7.			
TOTAL HOURS FOR SESSION			
TOTAL FORWARD			
TOTAL TO DATE			

FLIGHT MAP

POSTFLIGHT NOTES / JOURNAL ENTRIES:

I CERTIFY THAT THE FOREGOING ENTRIES ARE TRUE AND CORRECT:

PILOT: _____   SPOTTER: _____

UAS Pilot Log: Unmanned Aircraft Systems Logbook for Drone Pilots and Operators

FLIGHT ID / NO.: _____   Month _____ Day _____ Year _____

FLIGHT LOCATION: _____   Weather: _____

UNMANNED AIRCRAFT SYSTEM: ## CREW:

MANUFACTURER: _____   PILOT: _____

MODEL NUMBER: _____   SPOTTER: _____

PREFLIGHT CHECKLIST:

☐ Batteries Charged & Secure   ☐ Props OK & Tight   ☐ Compass Calibration
☐ Aircraft Hardware OK   ☐ Software / Firmware Update   ☐ Camera / FPV ON
☐ Equipment & Gear OK   ☐ Transmitter Control Power ON   ☐ Satellite Connection
☐ Transmitter Controls OK   ☐ Aircraft Power ON   ☐ Applications / Systems ON

PREFLIGHT NOTES:

SESSION FLIGHT INTERVALS	FLIGHT TIMES		
	START	STOP	TOTAL
1.			
2.			
3.			
4.			
5.			
6.			
7.			
TOTAL HOURS FOR SESSION			
TOTAL FORWARD			
TOTAL TO DATE			

FLIGHT MAP

POSTFLIGHT NOTES / JOURNAL ENTRIES:

I CERTIFY THAT THE FOREGOING ENTRIES ARE TRUE AND CORRECT:

PILOT: _____   SPOTTER: _____

UAS Pilot Log: Unmanned Aircraft Systems Logbook for Drone Pilots and Operators

FLIGHT ID / NO.: _____ Month _____ Day _____ Year _____

FLIGHT LOCATION: _____ Weather: _____

UNMANNED AIRCRAFT SYSTEM: **CREW:**

MANUFACTURER: _____   PILOT: _____

MODEL NUMBER: _____   SPOTTER: _____

PREFLIGHT CHECKLIST:
- ☐ Batteries Charged & Secure
- ☐ Aircraft Hardware OK
- ☐ Equipment & Gear OK
- ☐ Transmitter Controls OK
- ☐ Props OK & Tight
- ☐ Software / Firmware Update
- ☐ Transmitter Control Power ON
- ☐ Aircraft Power ON
- ☐ Compass Calibration
- ☐ Camera / FPV ON
- ☐ Satellite Connection
- ☐ Applications / Systems ON

PREFLIGHT NOTES:

SESSION FLIGHT INTERVALS	FLIGHT TIMES		
	START	STOP	TOTAL
1.			
2.			
3.			
4.			
5.			
6.			
7.			
TOTAL HOURS FOR SESSION			
TOTAL FORWARD			
TOTAL TO DATE			

FLIGHT MAP

POSTFLIGHT NOTES / JOURNAL ENTRIES:

I CERTIFY THAT THE FOREGOING ENTRIES ARE TRUE AND CORRECT:

PILOT: _____   SPOTTER: _____

UAS Pilot Log: Unmanned Aircraft Systems Logbook for Drone Pilots and Operators

FLIGHT ID / NO.: _____

Month _____ Day _____ Year _____

FLIGHT LOCATION: _____

Weather: _____

UNMANNED AIRCRAFT SYSTEM:

CREW:

MANUFACTURER: _____

PILOT: _____

MODEL NUMBER: _____

SPOTTER: _____

PREFLIGHT CHECKLIST:

- ☐ Batteries Charged & Secure
- ☐ Aircraft Hardware OK
- ☐ Equipment & Gear OK
- ☐ Transmitter Controls OK
- ☐ Props OK & Tight
- ☐ Software / Firmware Update
- ☐ Transmitter Control Power ON
- ☐ Aircraft Power ON
- ☐ Compass Calibration
- ☐ Camera / FPV ON
- ☐ Satellite Connection
- ☐ Applications / Systems ON

PREFLIGHT NOTES:

SESSION FLIGHT INTERVALS	FLIGHT TIMES		
	START	STOP	TOTAL
1.			
2.			
3.			
4.			
5.			
6.			
7.			
TOTAL HOURS FOR SESSION			
TOTAL FORWARD			
TOTAL TO DATE			

FLIGHT MAP

POSTFLIGHT NOTES / JOURNAL ENTRIES:

I CERTIFY THAT THE FOREGOING ENTRIES ARE TRUE AND CORRECT:

PILOT: _____

SPOTTER: _____

droneprep.com

UAS Pilot Log: Unmanned Aircraft Systems Logbook for Drone Pilots and Operators

FLIGHT ID / NO.: _____

Month _____ Day _____ Year _____

FLIGHT LOCATION: _____

Weather: _____

UNMANNED AIRCRAFT SYSTEM:

CREW:

MANUFACTURER: _____

PILOT: _____

MODEL NUMBER: _____

SPOTTER: _____

PREFLIGHT CHECKLIST:

- ☐ Batteries Charged & Secure
- ☐ Aircraft Hardware OK
- ☐ Equipment & Gear OK
- ☐ Transmitter Controls OK
- ☐ Props OK & Tight
- ☐ Software / Firmware Update
- ☐ Transmitter Control Power ON
- ☐ Aircraft Power ON
- ☐ Compass Calibration
- ☐ Camera / FPV ON
- ☐ Satellite Connection
- ☐ Applications / Systems ON

PREFLIGHT NOTES:

SESSION FLIGHT INTERVALS	FLIGHT TIMES		
	START	STOP	TOTAL
1.			
2.			
3.			
4.			
5.			
6.			
7.			
TOTAL HOURS FOR SESSION			
TOTAL FORWARD			
TOTAL TO DATE			

FLIGHT MAP

POSTFLIGHT NOTES / JOURNAL ENTRIES:

I CERTIFY THAT THE FOREGOING ENTRIES ARE TRUE AND CORRECT:

PILOT: _____

SPOTTER: _____

UAS Pilot Log: Unmanned Aircraft Systems Logbook for Drone Pilots and Operators

FLIGHT ID / NO.: _____ Month _____ Day _____ Year _____

FLIGHT LOCATION: _____ Weather: _____

UNMANNED AIRCRAFT SYSTEM: **CREW:**

MANUFACTURER: _____ PILOT: _____

MODEL NUMBER: _____ SPOTTER: _____

PREFLIGHT CHECKLIST:

- ☐ Batteries Charged & Secure
- ☐ Aircraft Hardware OK
- ☐ Equipment & Gear OK
- ☐ Transmitter Controls OK
- ☐ Props OK & Tight
- ☐ Software / Firmware Update
- ☐ Transmitter Control Power ON
- ☐ Aircraft Power ON
- ☐ Compass Calibration
- ☐ Camera / FPV ON
- ☐ Satellite Connection
- ☐ Applications / Systems ON

PREFLIGHT NOTES:

SESSION FLIGHT INTERVALS	FLIGHT TIMES		
	START	STOP	TOTAL
1.			
2.			
3.			
4.			
5.			
6.			
7.			
TOTAL HOURS FOR SESSION			
TOTAL FORWARD			
TOTAL TO DATE			

FLIGHT MAP

POSTFLIGHT NOTES / JOURNAL ENTRIES:

I CERTIFY THAT THE FOREGOING ENTRIES ARE TRUE AND CORRECT:

PILOT: _____ SPOTTER: _____

droneprep.com

UAS Pilot Log: Unmanned Aircraft Systems Logbook for Drone Pilots and Operators

FLIGHT ID / NO.: _____   Month _____ Day _____ Year _____

FLIGHT LOCATION: _____   Weather: _____

UNMANNED AIRCRAFT SYSTEM: **CREW:**

MANUFACTURER: _____   PILOT: _____

MODEL NUMBER: _____   SPOTTER: _____

PREFLIGHT CHECKLIST:
- ☐ Batteries Charged & Secure
- ☐ Aircraft Hardware OK
- ☐ Equipment & Gear OK
- ☐ Transmitter Controls OK
- ☐ Props OK & Tight
- ☐ Software / Firmware Update
- ☐ Transmitter Control Power ON
- ☐ Aircraft Power ON
- ☐ Compass Calibration
- ☐ Camera / FPV ON
- ☐ Satellite Connection
- ☐ Applications / Systems ON

PREFLIGHT NOTES:

SESSION FLIGHT INTERVALS	FLIGHT TIMES		
	START	STOP	TOTAL
1.			
2.			
3.			
4.			
5.			
6.			
7.			
TOTAL HOURS FOR SESSION			
TOTAL FORWARD			
TOTAL TO DATE			

FLIGHT MAP

POSTFLIGHT NOTES / JOURNAL ENTRIES:

I CERTIFY THAT THE FOREGOING ENTRIES ARE TRUE AND CORRECT:

PILOT: _____   SPOTTER: _____

UAS Pilot Log: Unmanned Aircraft Systems Logbook for Drone Pilots and Operators

FLIGHT ID / NO.: _____

Month _____ Day _____ Year _____

FLIGHT LOCATION: _____

Weather: _____

UNMANNED AIRCRAFT SYSTEM:

CREW:

MANUFACTURER: _____

PILOT: _____

MODEL NUMBER: _____

SPOTTER: _____

PREFLIGHT CHECKLIST:

- ☐ Batteries Charged & Secure
- ☐ Aircraft Hardware OK
- ☐ Equipment & Gear OK
- ☐ Transmitter Controls OK

- ☐ Props OK & Tight
- ☐ Software / Firmware Update
- ☐ Transmitter Control Power ON
- ☐ Aircraft Power ON

- ☐ Compass Calibration
- ☐ Camera / FPV ON
- ☐ Satellite Connection
- ☐ Applications / Systems ON

PREFLIGHT NOTES:

SESSION FLIGHT INTERVALS	FLIGHT TIMES		
	START	STOP	TOTAL
1.			
2.			
3.			
4.			
5.			
6.			
7.			
TOTAL HOURS FOR SESSION			
TOTAL FORWARD			
TOTAL TO DATE			

FLIGHT MAP

POSTFLIGHT NOTES / JOURNAL ENTRIES:

I CERTIFY THAT THE FOREGOING ENTRIES ARE TRUE AND CORRECT:

PILOT: _____

SPOTTER: _____

UAS Pilot Log: Unmanned Aircraft Systems Logbook for Drone Pilots and Operators

FLIGHT ID / NO.: _____

Month _____ Day _____ Year _____

FLIGHT LOCATION: _____

Weather: _____

UNMANNED AIRCRAFT SYSTEM:

CREW:

MANUFACTURER: _____

PILOT: _____

MODEL NUMBER: _____

SPOTTER: _____

PREFLIGHT CHECKLIST:

- ☐ Batteries Charged & Secure
- ☐ Aircraft Hardware OK
- ☐ Equipment & Gear OK
- ☐ Transmitter Controls OK
- ☐ Props OK & Tight
- ☐ Software / Firmware Update
- ☐ Transmitter Control Power ON
- ☐ Aircraft Power ON
- ☐ Compass Calibration
- ☐ Camera / FPV ON
- ☐ Satellite Connection
- ☐ Applications / Systems ON

PREFLIGHT NOTES:

SESSION FLIGHT INTERVALS	FLIGHT TIMES		
	START	STOP	TOTAL
1.			
2.			
3.			
4.			
5.			
6.			
7.			
TOTAL HOURS FOR SESSION			
TOTAL FORWARD			
TOTAL TO DATE			

FLIGHT MAP

POSTFLIGHT NOTES / JOURNAL ENTRIES:

I CERTIFY THAT THE FOREGOING ENTRIES ARE TRUE AND CORRECT:

PILOT: _____

SPOTTER: _____

UAS Pilot Log: Unmanned Aircraft Systems Logbook for Drone Pilots and Operators

FLIGHT ID / NO.: _____

Month _____ Day _____ Year _____

FLIGHT LOCATION: _____

Weather: _____

UNMANNED AIRCRAFT SYSTEM:

MANUFACTURER: _____

MODEL NUMBER: _____

CREW:

PILOT: _____

SPOTTER: _____

PREFLIGHT CHECKLIST:

- ☐ Batteries Charged & Secure
- ☐ Aircraft Hardware OK
- ☐ Equipment & Gear OK
- ☐ Transmitter Controls OK
- ☐ Props OK & Tight
- ☐ Software / Firmware Update
- ☐ Transmitter Control Power ON
- ☐ Aircraft Power ON
- ☐ Compass Calibration
- ☐ Camera / FPV ON
- ☐ Satellite Connection
- ☐ Applications / Systems ON

PREFLIGHT NOTES:

SESSION FLIGHT INTERVALS	FLIGHT TIMES		
	START	STOP	TOTAL
1.			
2.			
3.			
4.			
5.			
6.			
7.			
TOTAL HOURS FOR SESSION			
TOTAL FORWARD			
TOTAL TO DATE			

FLIGHT MAP

POSTFLIGHT NOTES / JOURNAL ENTRIES:

I CERTIFY THAT THE FOREGOING ENTRIES ARE TRUE AND CORRECT:

PILOT: _____   SPOTTER: _____

droneprep.com

UAS Pilot Log: Unmanned Aircraft Systems Logbook for Drone Pilots and Operators

FLIGHT ID / NO.: _____ Month _____ Day _____ Year _____

FLIGHT LOCATION: _____ Weather: _____

UNMANNED AIRCRAFT SYSTEM: **CREW:**

MANUFACTURER: _____ PILOT: _____

MODEL NUMBER: _____ SPOTTER: _____

PREFLIGHT CHECKLIST:

- ☐ Batteries Charged & Secure
- ☐ Aircraft Hardware OK
- ☐ Equipment & Gear OK
- ☐ Transmitter Controls OK
- ☐ Props OK & Tight
- ☐ Software / Firmware Update
- ☐ Transmitter Control Power ON
- ☐ Aircraft Power ON
- ☐ Compass Calibration
- ☐ Camera / FPV ON
- ☐ Satellite Connection
- ☐ Applications / Systems ON

PREFLIGHT NOTES:

SESSION FLIGHT INTERVALS	FLIGHT TIMES		
	START	STOP	TOTAL
1.			
2.			
3.			
4.			
5.			
6.			
7.			
TOTAL HOURS FOR SESSION			
TOTAL FORWARD			
TOTAL TO DATE			

FLIGHT MAP

POSTFLIGHT NOTES / JOURNAL ENTRIES:

I CERTIFY THAT THE FOREGOING ENTRIES ARE TRUE AND CORRECT:

PILOT: _____ SPOTTER: _____

UAS Pilot Log: Unmanned Aircraft Systems Logbook for Drone Pilots and Operators

FLIGHT ID / NO.: _____   Month _____ Day _____ Year _____

FLIGHT LOCATION: _____   Weather: _____

UNMANNED AIRCRAFT SYSTEM: **CREW:**

MANUFACTURER: _____   PILOT: _____

MODEL NUMBER: _____   SPOTTER: _____

PREFLIGHT CHECKLIST:

- ☐ Batteries Charged & Secure   ☐ Props OK & Tight   ☐ Compass Calibration
- ☐ Aircraft Hardware OK   ☐ Software / Firmware Update   ☐ Camera / FPV ON
- ☐ Equipment & Gear OK   ☐ Transmitter Control Power ON ☐ Satellite Connection
- ☐ Transmitter Controls OK   ☐ Aircraft Power ON   ☐ Applications / Systems ON

PREFLIGHT NOTES:

SESSION FLIGHT INTERVALS	FLIGHT TIMES		
	START	STOP	TOTAL
1.			
2.			
3.			
4.			
5.			
6.			
7.			
TOTAL HOURS FOR SESSION			
TOTAL FORWARD			
TOTAL TO DATE			

FLIGHT MAP

POSTFLIGHT NOTES / JOURNAL ENTRIES:

I CERTIFY THAT THE FOREGOING ENTRIES ARE TRUE AND CORRECT:

PILOT: _____   SPOTTER: _____

droneprep.com

UAS Pilot Log: Unmanned Aircraft Systems Logbook for Drone Pilots and Operators

FLIGHT ID / NO.: _____

Month _____ Day _____ Year _____

FLIGHT LOCATION: _____

Weather: _____

UNMANNED AIRCRAFT SYSTEM:

CREW:

MANUFACTURER: _____

PILOT: _____

MODEL NUMBER: _____

SPOTTER: _____

PREFLIGHT CHECKLIST:
- ☐ Batteries Charged & Secure
- ☐ Aircraft Hardware OK
- ☐ Equipment & Gear OK
- ☐ Transmitter Controls OK
- ☐ Props OK & Tight
- ☐ Software / Firmware Update
- ☐ Transmitter Control Power ON
- ☐ Aircraft Power ON
- ☐ Compass Calibration
- ☐ Camera / FPV ON
- ☐ Satellite Connection
- ☐ Applications / Systems ON

PREFLIGHT NOTES:

SESSION FLIGHT INTERVALS	FLIGHT TIMES		
	START	STOP	TOTAL
1.			
2.			
3.			
4.			
5.			
6.			
7.			
TOTAL HOURS FOR SESSION			
TOTAL FORWARD			
TOTAL TO DATE			

FLIGHT MAP

POSTFLIGHT NOTES / JOURNAL ENTRIES:

I CERTIFY THAT THE FOREGOING ENTRIES ARE TRUE AND CORRECT:

PILOT: _____ SPOTTER: _____

UAS Pilot Log: Unmanned Aircraft Systems Logbook for Drone Pilots and Operators

FLIGHT ID / NO.: _____   Month _____ Day _____ Year _____

FLIGHT LOCATION: _____   Weather: _____

UNMANNED AIRCRAFT SYSTEM: **CREW:**

MANUFACTURER: _____   PILOT: _____

MODEL NUMBER: _____   SPOTTER: _____

PREFLIGHT CHECKLIST:

- ☐ Batteries Charged & Secure   ☐ Props OK & Tight   ☐ Compass Calibration
- ☐ Aircraft Hardware OK   ☐ Software / Firmware Update   ☐ Camera / FPV ON
- ☐ Equipment & Gear OK   ☐ Transmitter Control Power ON   ☐ Satellite Connection
- ☐ Transmitter Controls OK   ☐ Aircraft Power ON   ☐ Applications / Systems ON

PREFLIGHT NOTES:

SESSION FLIGHT INTERVALS	FLIGHT TIMES		
	START	STOP	TOTAL
1.			
2.			
3.			
4.			
5.			
6.			
7.			
TOTAL HOURS FOR SESSION			
TOTAL FORWARD			
TOTAL TO DATE			

FLIGHT MAP

POSTFLIGHT NOTES / JOURNAL ENTRIES:

I CERTIFY THAT THE FOREGOING ENTRIES ARE TRUE AND CORRECT:

PILOT: _____   SPOTTER: _____

UAS Pilot Log: Unmanned Aircraft Systems Logbook for Drone Pilots and Operators

FLIGHT ID / NO.: _____   Month _____ Day _____ Year _____

FLIGHT LOCATION: _____   Weather: _____

UNMANNED AIRCRAFT SYSTEM: **CREW:**

MANUFACTURER: _____   PILOT: _____

MODEL NUMBER: _____   SPOTTER: _____

PREFLIGHT CHECKLIST:

- ☐ Batteries Charged & Secure   ☐ Props OK & Tight   ☐ Compass Calibration
- ☐ Aircraft Hardware OK   ☐ Software / Firmware Update   ☐ Camera / FPV ON
- ☐ Equipment & Gear OK   ☐ Transmitter Control Power ON   ☐ Satellite Connection
- ☐ Transmitter Controls OK   ☐ Aircraft Power ON   ☐ Applications / Systems ON

PREFLIGHT NOTES:

SESSION FLIGHT INTERVALS	FLIGHT TIMES		
	START	STOP	TOTAL
1.			
2.			
3.			
4.			
5.			
6.			
7.			
TOTAL HOURS FOR SESSION			
TOTAL FORWARD			
TOTAL TO DATE			

FLIGHT MAP

POSTFLIGHT NOTES / JOURNAL ENTRIES:

I CERTIFY THAT THE FOREGOING ENTRIES ARE TRUE AND CORRECT:

PILOT: _____   SPOTTER: _____

UAS Pilot Log: Unmanned Aircraft Systems Logbook for Drone Pilots and Operators

FLIGHT ID / NO.: _____

Month _____ Day _____ Year _____

FLIGHT LOCATION: _____

Weather: _____

UNMANNED AIRCRAFT SYSTEM:

CREW:

MANUFACTURER: _____

PILOT: _____

MODEL NUMBER: _____

SPOTTER: _____

PREFLIGHT CHECKLIST:

- ☐ Batteries Charged & Secure
- ☐ Aircraft Hardware OK
- ☐ Equipment & Gear OK
- ☐ Transmitter Controls OK
- ☐ Props OK & Tight
- ☐ Software / Firmware Update
- ☐ Transmitter Control Power ON
- ☐ Aircraft Power ON
- ☐ Compass Calibration
- ☐ Camera / FPV ON
- ☐ Satellite Connection
- ☐ Applications / Systems ON

PREFLIGHT NOTES:

SESSION FLIGHT INTERVALS	FLIGHT TIMES		
	START	STOP	TOTAL
1.			
2.			
3.			
4.			
5.			
6.			
7.			
TOTAL HOURS FOR SESSION			
TOTAL FORWARD			
TOTAL TO DATE			

FLIGHT MAP

POSTFLIGHT NOTES / JOURNAL ENTRIES:

I CERTIFY THAT THE FOREGOING ENTRIES ARE TRUE AND CORRECT:

PILOT: _____

SPOTTER: _____

droneprep.com

UAS Pilot Log: Unmanned Aircraft Systems Logbook for Drone Pilots and Operators

FLIGHT ID / NO.: _____

Month _____ Day _____ Year _____

FLIGHT LOCATION: _____

Weather: _____

UNMANNED AIRCRAFT SYSTEM:

CREW:

MANUFACTURER: _____

PILOT: _____

MODEL NUMBER: _____

SPOTTER: _____

PREFLIGHT CHECKLIST:
- ☐ Batteries Charged & Secure
- ☐ Aircraft Hardware OK
- ☐ Equipment & Gear OK
- ☐ Transmitter Controls OK
- ☐ Props OK & Tight
- ☐ Software / Firmware Update
- ☐ Transmitter Control Power ON
- ☐ Aircraft Power ON
- ☐ Compass Calibration
- ☐ Camera / FPV ON
- ☐ Satellite Connection
- ☐ Applications / Systems ON

PREFLIGHT NOTES:

SESSION FLIGHT INTERVALS	FLIGHT TIMES		
	START	STOP	TOTAL
1.			
2.			
3.			
4.			
5.			
6.			
7.			
TOTAL HOURS FOR SESSION			
TOTAL FORWARD			
TOTAL TO DATE			

FLIGHT MAP

POSTFLIGHT NOTES / JOURNAL ENTRIES:

I CERTIFY THAT THE FOREGOING ENTRIES ARE TRUE AND CORRECT:

PILOT: _____

SPOTTER: _____

UAS Pilot Log: Unmanned Aircraft Systems Logbook for Drone Pilots and Operators

FLIGHT ID / NO.: _____   Month _____ Day _____ Year _____

FLIGHT LOCATION: _____   Weather: _____

UNMANNED AIRCRAFT SYSTEM: **CREW:**

MANUFACTURER: _____   PILOT: _____

MODEL NUMBER: _____   SPOTTER: _____

PREFLIGHT CHECKLIST:

- ☐ Batteries Charged & Secure
- ☐ Aircraft Hardware OK
- ☐ Equipment & Gear OK
- ☐ Transmitter Controls OK
- ☐ Props OK & Tight
- ☐ Software / Firmware Update
- ☐ Transmitter Control Power ON
- ☐ Aircraft Power ON
- ☐ Compass Calibration
- ☐ Camera / FPV ON
- ☐ Satellite Connection
- ☐ Applications / Systems ON

PREFLIGHT NOTES:

_____

SESSION FLIGHT INTERVALS	FLIGHT TIMES		
	START	STOP	TOTAL
1.			
2.			
3.			
4.			
5.			
6.			
7.			
TOTAL HOURS FOR SESSION			
TOTAL FORWARD			
TOTAL TO DATE			

FLIGHT MAP

POSTFLIGHT NOTES / JOURNAL ENTRIES:

I CERTIFY THAT THE FOREGOING ENTRIES ARE TRUE AND CORRECT:

PILOT: _____   SPOTTER: _____

droneprep.com

UAS Pilot Log: Unmanned Aircraft Systems Logbook for Drone Pilots and Operators

FLIGHT ID / NO.: _____ Month _____ Day _____ Year _____

FLIGHT LOCATION: _____ Weather: _____

UNMANNED AIRCRAFT SYSTEM: **CREW:**

MANUFACTURER: _____ PILOT: _____

MODEL NUMBER: _____ SPOTTER: _____

PREFLIGHT CHECKLIST:
- ☐ Batteries Charged & Secure   ☐ Props OK & Tight   ☐ Compass Calibration
- ☐ Aircraft Hardware OK   ☐ Software / Firmware Update   ☐ Camera / FPV ON
- ☐ Equipment & Gear OK   ☐ Transmitter Control Power ON   ☐ Satellite Connection
- ☐ Transmitter Controls OK   ☐ Aircraft Power ON   ☐ Applications / Systems ON

PREFLIGHT NOTES:

SESSION FLIGHT INTERVALS	FLIGHT TIMES		
	START	STOP	TOTAL
1.			
2.			
3.			
4.			
5.			
6.			
7.			
TOTAL HOURS FOR SESSION			
TOTAL FORWARD			
TOTAL TO DATE			

FLIGHT MAP

POSTFLIGHT NOTES / JOURNAL ENTRIES:

I CERTIFY THAT THE FOREGOING ENTRIES ARE TRUE AND CORRECT:

PILOT: _____   SPOTTER: _____

UAS Pilot Log: Unmanned Aircraft Systems Logbook for Drone Pilots and Operators

FLIGHT ID / NO.: _____   Month _____ Day _____ Year _____

FLIGHT LOCATION: _____   Weather: _____

UNMANNED AIRCRAFT SYSTEM: **CREW:**

MANUFACTURER: _____   PILOT: _____

MODEL NUMBER: _____   SPOTTER: _____

PREFLIGHT CHECKLIST:

- ☐ Batteries Charged & Secure
- ☐ Aircraft Hardware OK
- ☐ Equipment & Gear OK
- ☐ Transmitter Controls OK
- ☐ Props OK & Tight
- ☐ Software / Firmware Update
- ☐ Transmitter Control Power ON
- ☐ Aircraft Power ON
- ☐ Compass Calibration
- ☐ Camera / FPV ON
- ☐ Satellite Connection
- ☐ Applications / Systems ON

PREFLIGHT NOTES:

SESSION FLIGHT INTERVALS	FLIGHT TIMES		
	START	STOP	TOTAL
1.			
2.			
3.			
4.			
5.			
6.			
7.			
TOTAL HOURS FOR SESSION			
TOTAL FORWARD			
TOTAL TO DATE			

FLIGHT MAP

POSTFLIGHT NOTES / JOURNAL ENTRIES:

I CERTIFY THAT THE FOREGOING ENTRIES ARE TRUE AND CORRECT:

PILOT: _____   SPOTTER: _____

droneprep.com

UAS Pilot Log: Unmanned Aircraft Systems Logbook for Drone Pilots and Operators

FLIGHT ID / NO.: _____   Month _____ Day _____ Year _____

FLIGHT LOCATION: _____   Weather: _____

UNMANNED AIRCRAFT SYSTEM: **CREW:**

MANUFACTURER: _____   PILOT: _____

MODEL NUMBER: _____   SPOTTER: _____

PREFLIGHT CHECKLIST:

- ☐ Batteries Charged & Secure
- ☐ Aircraft Hardware OK
- ☐ Equipment & Gear OK
- ☐ Transmitter Controls OK
- ☐ Props OK & Tight
- ☐ Software / Firmware Update
- ☐ Transmitter Control Power ON
- ☐ Aircraft Power ON
- ☐ Compass Calibration
- ☐ Camera / FPV ON
- ☐ Satellite Connection
- ☐ Applications / Systems ON

PREFLIGHT NOTES:

_____

SESSION FLIGHT INTERVALS	FLIGHT TIMES		
	START	STOP	TOTAL
1.			
2.			
3.			
4.			
5.			
6.			
7.			
TOTAL HOURS FOR SESSION			
TOTAL FORWARD			
TOTAL TO DATE			

FLIGHT MAP

POSTFLIGHT NOTES / JOURNAL ENTRIES:

I CERTIFY THAT THE FOREGOING ENTRIES ARE TRUE AND CORRECT:

PILOT: _____   SPOTTER: _____

UAS Pilot Log: Unmanned Aircraft Systems Logbook for Drone Pilots and Operators

FLIGHT ID / NO.: _____

Month _____ Day _____ Year _____

FLIGHT LOCATION: _____

Weather: _____

UNMANNED AIRCRAFT SYSTEM:

CREW:

MANUFACTURER: _____

PILOT: _____

MODEL NUMBER: _____

SPOTTER: _____

PREFLIGHT CHECKLIST:

- ☐ Batteries Charged & Secure
- ☐ Aircraft Hardware OK
- ☐ Equipment & Gear OK
- ☐ Transmitter Controls OK

- ☐ Props OK & Tight
- ☐ Software / Firmware Update
- ☐ Transmitter Control Power ON
- ☐ Aircraft Power ON

- ☐ Compass Calibration
- ☐ Camera / FPV ON
- ☐ Satellite Connection
- ☐ Applications / Systems ON

PREFLIGHT NOTES:

SESSION FLIGHT INTERVALS	FLIGHT TIMES		
	START	STOP	TOTAL
1.			
2.			
3.			
4.			
5.			
6.			
7.			
TOTAL HOURS FOR SESSION			
TOTAL FORWARD			
TOTAL TO DATE			

FLIGHT MAP

POSTFLIGHT NOTES / JOURNAL ENTRIES:

I CERTIFY THAT THE FOREGOING ENTRIES ARE TRUE AND CORRECT:

PILOT: _____

SPOTTER: _____

UAS Pilot Log: Unmanned Aircraft Systems Logbook for Drone Pilots and Operators

FLIGHT ID / NO.: _____   Month _____ Day _____ Year _____

FLIGHT LOCATION: _____   Weather: _____

UNMANNED AIRCRAFT SYSTEM: **CREW:**

MANUFACTURER: _____   PILOT: _____

MODEL NUMBER: _____   SPOTTER: _____

PREFLIGHT CHECKLIST:
- ☐ Batteries Charged & Secure
- ☐ Aircraft Hardware OK
- ☐ Equipment & Gear OK
- ☐ Transmitter Controls OK
- ☐ Props OK & Tight
- ☐ Software / Firmware Update
- ☐ Transmitter Control Power ON
- ☐ Aircraft Power ON
- ☐ Compass Calibration
- ☐ Camera / FPV ON
- ☐ Satellite Connection
- ☐ Applications / Systems ON

PREFLIGHT NOTES:

SESSION FLIGHT INTERVALS	FLIGHT TIMES		
	START	STOP	TOTAL
1.			
2.			
3.			
4.			
5.			
6.			
7.			
TOTAL HOURS FOR SESSION			
TOTAL FORWARD			
TOTAL TO DATE			

FLIGHT MAP

POSTFLIGHT NOTES / JOURNAL ENTRIES:

I CERTIFY THAT THE FOREGOING ENTRIES ARE TRUE AND CORRECT:

PILOT: _____   SPOTTER: _____

UAS Pilot Log: Unmanned Aircraft Systems Logbook for Drone Pilots and Operators

FLIGHT ID / NO.: _____	Month _____ Day _____ Year _____

FLIGHT LOCATION: _____	Weather: _____

UNMANNED AIRCRAFT SYSTEM:	**CREW:**

MANUFACTURER: _____	PILOT: _____

MODEL NUMBER: _____	SPOTTER: _____

PREFLIGHT CHECKLIST:

- ☐ Batteries Charged & Secure
- ☐ Aircraft Hardware OK
- ☐ Equipment & Gear OK
- ☐ Transmitter Controls OK
- ☐ Props OK & Tight
- ☐ Software / Firmware Update
- ☐ Transmitter Control Power ON
- ☐ Aircraft Power ON
- ☐ Compass Calibration
- ☐ Camera / FPV ON
- ☐ Satellite Connection
- ☐ Applications / Systems ON

PREFLIGHT NOTES:

SESSION FLIGHT INTERVALS	FLIGHT TIMES		
	START	STOP	TOTAL
1.			
2.			
3.			
4.			
5.			
6.			
7.			
TOTAL HOURS FOR SESSION			
TOTAL FORWARD			
TOTAL TO DATE			

FLIGHT MAP

POSTFLIGHT NOTES / JOURNAL ENTRIES:

I CERTIFY THAT THE FOREGOING ENTRIES ARE TRUE AND CORRECT:

PILOT: _____	SPOTTER: _____

UAS Pilot Log: Unmanned Aircraft Systems Logbook for Drone Pilots and Operators

FLIGHT ID / NO.: _____   Month _____ Day _____ Year _____

FLIGHT LOCATION: _____   Weather: _____

UNMANNED AIRCRAFT SYSTEM: **CREW:**

MANUFACTURER: _____   PILOT: _____

MODEL NUMBER: _____   SPOTTER: _____

PREFLIGHT CHECKLIST:

- ☐ Batteries Charged & Secure
- ☐ Aircraft Hardware OK
- ☐ Equipment & Gear OK
- ☐ Transmitter Controls OK
- ☐ Props OK & Tight
- ☐ Software / Firmware Update
- ☐ Transmitter Control Power ON
- ☐ Aircraft Power ON
- ☐ Compass Calibration
- ☐ Camera / FPV ON
- ☐ Satellite Connection
- ☐ Applications / Systems ON

PREFLIGHT NOTES:

SESSION FLIGHT INTERVALS	FLIGHT TIMES		
	START	STOP	TOTAL
1.			
2.			
3.			
4.			
5.			
6.			
7.			
TOTAL HOURS FOR SESSION			
TOTAL FORWARD			
TOTAL TO DATE			

FLIGHT MAP

POSTFLIGHT NOTES / JOURNAL ENTRIES:

I CERTIFY THAT THE FOREGOING ENTRIES ARE TRUE AND CORRECT:

PILOT: _____   SPOTTER: _____

UAS Pilot Log: Unmanned Aircraft Systems Logbook for Drone Pilots and Operators

FLIGHT ID / NO.: _____

Month _____ Day _____ Year _____

FLIGHT LOCATION: _____

Weather: _____

UNMANNED AIRCRAFT SYSTEM:

CREW:

MANUFACTURER: _____

PILOT: _____

MODEL NUMBER: _____

SPOTTER: _____

PREFLIGHT CHECKLIST:

- ☐ Batteries Charged & Secure
- ☐ Aircraft Hardware OK
- ☐ Equipment & Gear OK
- ☐ Transmitter Controls OK

- ☐ Props OK & Tight
- ☐ Software / Firmware Update
- ☐ Transmitter Control Power ON
- ☐ Aircraft Power ON

- ☐ Compass Calibration
- ☐ Camera / FPV ON
- ☐ Satellite Connection
- ☐ Applications / Systems ON

PREFLIGHT NOTES:

SESSION FLIGHT INTERVALS	FLIGHT TIMES		
	START	STOP	TOTAL
1.			
2.			
3.			
4.			
5.			
6.			
7.			
TOTAL HOURS FOR SESSION			
TOTAL FORWARD			
TOTAL TO DATE			

FLIGHT MAP

POSTFLIGHT NOTES / JOURNAL ENTRIES:

I CERTIFY THAT THE FOREGOING ENTRIES ARE TRUE AND CORRECT:

PILOT: _____

SPOTTER: _____

droneprep.com

UAS Pilot Log: Unmanned Aircraft Systems Logbook for Drone Pilots and Operators

FLIGHT ID / NO.: _____

Month _____ Day _____ Year _____

FLIGHT LOCATION: _____

Weather: _____

UNMANNED AIRCRAFT SYSTEM:

CREW:

MANUFACTURER: _____

PILOT: _____

MODEL NUMBER: _____

SPOTTER: _____

PREFLIGHT CHECKLIST:
- ☐ Batteries Charged & Secure
- ☐ Aircraft Hardware OK
- ☐ Equipment & Gear OK
- ☐ Transmitter Controls OK
- ☐ Props OK & Tight
- ☐ Software / Firmware Update
- ☐ Transmitter Control Power ON
- ☐ Aircraft Power ON
- ☐ Compass Calibration
- ☐ Camera / FPV ON
- ☐ Satellite Connection
- ☐ Applications / Systems ON

PREFLIGHT NOTES:

SESSION FLIGHT INTERVALS	FLIGHT TIMES		
	START	STOP	TOTAL
1.			
2.			
3.			
4.			
5.			
6.			
7.			
TOTAL HOURS FOR SESSION			
TOTAL FORWARD			
TOTAL TO DATE			

FLIGHT MAP

POSTFLIGHT NOTES / JOURNAL ENTRIES:

I CERTIFY THAT THE FOREGOING ENTRIES ARE TRUE AND CORRECT:

PILOT: _____   SPOTTER: _____

droneprep.com

UAS Pilot Log: Unmanned Aircraft Systems Logbook for Drone Pilots and Operators

FLIGHT ID / NO.: _____

Month _____ Day _____ Year _____

FLIGHT LOCATION: _____

Weather: _____

UNMANNED AIRCRAFT SYSTEM:

CREW:

MANUFACTURER: _____

PILOT: _____

MODEL NUMBER: _____

SPOTTER: _____

PREFLIGHT CHECKLIST:

- ☐ Batteries Charged & Secure
- ☐ Aircraft Hardware OK
- ☐ Equipment & Gear OK
- ☐ Transmitter Controls OK
- ☐ Props OK & Tight
- ☐ Software / Firmware Update
- ☐ Transmitter Control Power ON
- ☐ Aircraft Power ON
- ☐ Compass Calibration
- ☐ Camera / FPV ON
- ☐ Satellite Connection
- ☐ Applications / Systems ON

PREFLIGHT NOTES:

SESSION FLIGHT INTERVALS	FLIGHT TIMES		
	START	STOP	TOTAL
1.			
2.			
3.			
4.			
5.			
6.			
7.			
TOTAL HOURS FOR SESSION			
TOTAL FORWARD			
TOTAL TO DATE			

FLIGHT MAP

POSTFLIGHT NOTES / JOURNAL ENTRIES:

I CERTIFY THAT THE FOREGOING ENTRIES ARE TRUE AND CORRECT:

PILOT: _____

SPOTTER: _____

droneprep.com

UAS Pilot Log: Unmanned Aircraft Systems Logbook for Drone Pilots and Operators

FLIGHT ID / NO.: _____   Month _____ Day _____ Year _____

FLIGHT LOCATION: _____   Weather: _____

UNMANNED AIRCRAFT SYSTEM: **CREW:**

MANUFACTURER: _____   PILOT: _____

MODEL NUMBER: _____   SPOTTER: _____

PREFLIGHT CHECKLIST:
- ☐ Batteries Charged & Secure   ☐ Props OK & Tight   ☐ Compass Calibration
- ☐ Aircraft Hardware OK   ☐ Software / Firmware Update   ☐ Camera / FPV ON
- ☐ Equipment & Gear OK   ☐ Transmitter Control Power ON   ☐ Satellite Connection
- ☐ Transmitter Controls OK   ☐ Aircraft Power ON   ☐ Applications / Systems ON

PREFLIGHT NOTES:

SESSION FLIGHT INTERVALS	FLIGHT TIMES		
	START	STOP	TOTAL
1.			
2.			
3.			
4.			
5.			
6.			
7.			
TOTAL HOURS FOR SESSION			
TOTAL FORWARD			
TOTAL TO DATE			

FLIGHT MAP

POSTFLIGHT NOTES / JOURNAL ENTRIES:

I CERTIFY THAT THE FOREGOING ENTRIES ARE TRUE AND CORRECT:

PILOT: _____   SPOTTER: _____

UAS Pilot Log: Unmanned Aircraft Systems Logbook for Drone Pilots and Operators

FLIGHT ID / NO.: _____   Month _____ Day _____ Year _____

FLIGHT LOCATION: _____   Weather: _____

UNMANNED AIRCRAFT SYSTEM: **CREW:**

MANUFACTURER: _____   PILOT: _____

MODEL NUMBER: _____   SPOTTER: _____

PREFLIGHT CHECKLIST:

- ☐ Batteries Charged & Secure
- ☐ Aircraft Hardware OK
- ☐ Equipment & Gear OK
- ☐ Transmitter Controls OK
- ☐ Props OK & Tight
- ☐ Software / Firmware Update
- ☐ Transmitter Control Power ON
- ☐ Aircraft Power ON
- ☐ Compass Calibration
- ☐ Camera / FPV ON
- ☐ Satellite Connection
- ☐ Applications / Systems ON

PREFLIGHT NOTES:

SESSION FLIGHT INTERVALS	FLIGHT TIMES		
	START	STOP	TOTAL
1.			
2.			
3.			
4.			
5.			
6.			
7.			
TOTAL HOURS FOR SESSION			
TOTAL FORWARD			
TOTAL TO DATE			

FLIGHT MAP

POSTFLIGHT NOTES / JOURNAL ENTRIES:

I CERTIFY THAT THE FOREGOING ENTRIES ARE TRUE AND CORRECT:

PILOT: _____   SPOTTER: _____

droneprep.com

UAS Pilot Log: Unmanned Aircraft Systems Logbook for Drone Pilots and Operators

FLIGHT ID / NO.: _____

Month _____ Day _____ Year _____

FLIGHT LOCATION: _____

Weather: _____

UNMANNED AIRCRAFT SYSTEM:

CREW:

MANUFACTURER: _____

PILOT: _____

MODEL NUMBER: _____

SPOTTER: _____

PREFLIGHT CHECKLIST:
- ☐ Batteries Charged & Secure
- ☐ Aircraft Hardware OK
- ☐ Equipment & Gear OK
- ☐ Transmitter Controls OK
- ☐ Props OK & Tight
- ☐ Software / Firmware Update
- ☐ Transmitter Control Power ON
- ☐ Aircraft Power ON
- ☐ Compass Calibration
- ☐ Camera / FPV ON
- ☐ Satellite Connection
- ☐ Applications / Systems ON

PREFLIGHT NOTES:

SESSION FLIGHT INTERVALS	FLIGHT TIMES		
	START	STOP	TOTAL
1.			
2.			
3.			
4.			
5.			
6.			
7.			
TOTAL HOURS FOR SESSION			
TOTAL FORWARD			
TOTAL TO DATE			

FLIGHT MAP

POSTFLIGHT NOTES / JOURNAL ENTRIES:

I CERTIFY THAT THE FOREGOING ENTRIES ARE TRUE AND CORRECT:

PILOT: _____

SPOTTER: _____

droneprep.com

UAS Pilot Log: Unmanned Aircraft Systems Logbook for Drone Pilots and Operators

FLIGHT ID / NO.: _____

Month _____ Day _____ Year _____

FLIGHT LOCATION: _____

Weather: _____

UNMANNED AIRCRAFT SYSTEM:

CREW:

MANUFACTURER: _____

PILOT: _____

MODEL NUMBER: _____

SPOTTER: _____

PREFLIGHT CHECKLIST:

- ☐ Batteries Charged & Secure
- ☐ Aircraft Hardware OK
- ☐ Equipment & Gear OK
- ☐ Transmitter Controls OK
- ☐ Props OK & Tight
- ☐ Software / Firmware Update
- ☐ Transmitter Control Power ON
- ☐ Aircraft Power ON
- ☐ Compass Calibration
- ☐ Camera / FPV ON
- ☐ Satellite Connection
- ☐ Applications / Systems ON

PREFLIGHT NOTES:

SESSION FLIGHT INTERVALS	FLIGHT TIMES		
	START	STOP	TOTAL
1.			
2.			
3.			
4.			
5.			
6.			
7.			
TOTAL HOURS FOR SESSION			
TOTAL FORWARD			
TOTAL TO DATE			

FLIGHT MAP

POSTFLIGHT NOTES / JOURNAL ENTRIES:

I CERTIFY THAT THE FOREGOING ENTRIES ARE TRUE AND CORRECT:

PILOT: _____   SPOTTER: _____

droneprep.com

UAS Pilot Log: Unmanned Aircraft Systems Logbook for Drone Pilots and Operators

FLIGHT ID / NO.: _____

Month _____ Day _____ Year _____

FLIGHT LOCATION: _____

Weather: _____

UNMANNED AIRCRAFT SYSTEM:

MANUFACTURER: _____

MODEL NUMBER: _____

CREW:

PILOT: _____

SPOTTER: _____

PREFLIGHT CHECKLIST:

- ☐ Batteries Charged & Secure
- ☐ Aircraft Hardware OK
- ☐ Equipment & Gear OK
- ☐ Transmitter Controls OK

- ☐ Props OK & Tight
- ☐ Software / Firmware Update
- ☐ Transmitter Control Power ON
- ☐ Aircraft Power ON

- ☐ Compass Calibration
- ☐ Camera / FPV ON
- ☐ Satellite Connection
- ☐ Applications / Systems ON

PREFLIGHT NOTES:

_____

SESSION FLIGHT INTERVALS	FLIGHT TIMES		
	START	STOP	TOTAL
1.			
2.			
3.			
4.			
5.			
6.			
7.			
TOTAL HOURS FOR SESSION			
TOTAL FORWARD			
TOTAL TO DATE			

FLIGHT MAP

POSTFLIGHT NOTES / JOURNAL ENTRIES:

I CERTIFY THAT THE FOREGOING ENTRIES ARE TRUE AND CORRECT:

PILOT: _____ SPOTTER: _____

droneprep.com

UAS Pilot Log: Unmanned Aircraft Systems Logbook for Drone Pilots and Operators

FLIGHT ID / NO.: _____

Month _____ Day _____ Year _____

FLIGHT LOCATION: _____

Weather: _____

UNMANNED AIRCRAFT SYSTEM:

CREW:

MANUFACTURER: _____

PILOT: _____

MODEL NUMBER: _____

SPOTTER: _____

PREFLIGHT CHECKLIST:

- ☐ Batteries Charged & Secure
- ☐ Aircraft Hardware OK
- ☐ Equipment & Gear OK
- ☐ Transmitter Controls OK

- ☐ Props OK & Tight
- ☐ Software / Firmware Update
- ☐ Transmitter Control Power ON
- ☐ Aircraft Power ON

- ☐ Compass Calibration
- ☐ Camera / FPV ON
- ☐ Satellite Connection
- ☐ Applications / Systems ON

PREFLIGHT NOTES:

SESSION FLIGHT INTERVALS	FLIGHT TIMES		
	START	STOP	TOTAL
1.			
2.			
3.			
4.			
5.			
6.			
7.			
TOTAL HOURS FOR SESSION			
TOTAL FORWARD			
TOTAL TO DATE			

FLIGHT MAP

POSTFLIGHT NOTES / JOURNAL ENTRIES:

I CERTIFY THAT THE FOREGOING ENTRIES ARE TRUE AND CORRECT:

PILOT: _____

SPOTTER: _____

UAS Pilot Log: Unmanned Aircraft Systems Logbook for Drone Pilots and Operators

FLIGHT ID / NO.: _____　　　Month _____ Day _____ Year _____

FLIGHT LOCATION: _____　　　Weather: _____

UNMANNED AIRCRAFT SYSTEM:　　　　　　　　**CREW:**

MANUFACTURER: _____　　PILOT: _____

MODEL NUMBER: _____　　SPOTTER: _____

PREFLIGHT CHECKLIST:
- ☐ Batteries Charged & Secure　　☐ Props OK & Tight　　☐ Compass Calibration
- ☐ Aircraft Hardware OK　　☐ Software / Firmware Update　　☐ Camera / FPV ON
- ☐ Equipment & Gear OK　　☐ Transmitter Control Power ON　　☐ Satellite Connection
- ☐ Transmitter Controls OK　　☐ Aircraft Power ON　　☐ Applications / Systems ON

PREFLIGHT NOTES:

_____

SESSION FLIGHT INTERVALS	FLIGHT TIMES		
	START	STOP	TOTAL
1.			
2.			
3.			
4.			
5.			
6.			
7.			
TOTAL HOURS FOR SESSION			
TOTAL FORWARD			
TOTAL TO DATE			

FLIGHT MAP

POSTFLIGHT NOTES / JOURNAL ENTRIES:

I CERTIFY THAT THE FOREGOING ENTRIES ARE TRUE AND CORRECT:

PILOT: _____　　SPOTTER: _____

UAS Pilot Log: Unmanned Aircraft Systems Logbook for Drone Pilots and Operators

FLIGHT ID / NO.: _____   Month _____ Day _____ Year _____

FLIGHT LOCATION: _____   Weather: _____

UNMANNED AIRCRAFT SYSTEM: **CREW:**

MANUFACTURER: _____   PILOT: _____

MODEL NUMBER: _____   SPOTTER: _____

PREFLIGHT CHECKLIST:

- ☐ Batteries Charged & Secure
- ☐ Aircraft Hardware OK
- ☐ Equipment & Gear OK
- ☐ Transmitter Controls OK
- ☐ Props OK & Tight
- ☐ Software / Firmware Update
- ☐ Transmitter Control Power ON
- ☐ Aircraft Power ON
- ☐ Compass Calibration
- ☐ Camera / FPV ON
- ☐ Satellite Connection
- ☐ Applications / Systems ON

PREFLIGHT NOTES:

SESSION FLIGHT INTERVALS	FLIGHT TIMES		
	START	STOP	TOTAL
1.			
2.			
3.			
4.			
5.			
6.			
7.			
TOTAL HOURS FOR SESSION			
TOTAL FORWARD			
TOTAL TO DATE			

FLIGHT MAP

POSTFLIGHT NOTES / JOURNAL ENTRIES:

I CERTIFY THAT THE FOREGOING ENTRIES ARE TRUE AND CORRECT:

PILOT: _____   SPOTTER: _____

UAS Pilot Log: Unmanned Aircraft Systems Logbook for Drone Pilots and Operators

FLIGHT ID / NO.: _____   Month _____ Day _____ Year _____

FLIGHT LOCATION: _____   Weather: _____

UNMANNED AIRCRAFT SYSTEM: **CREW:**

MANUFACTURER: _____   PILOT: _____

MODEL NUMBER: _____   SPOTTER: _____

PREFLIGHT CHECKLIST:
- ☐ Batteries Charged & Secure   ☐ Props OK & Tight   ☐ Compass Calibration
- ☐ Aircraft Hardware OK   ☐ Software / Firmware Update   ☐ Camera / FPV ON
- ☐ Equipment & Gear OK   ☐ Transmitter Control Power ON   ☐ Satellite Connection
- ☐ Transmitter Controls OK   ☐ Aircraft Power ON   ☐ Applications / Systems ON

PREFLIGHT NOTES:

SESSION FLIGHT INTERVALS	FLIGHT TIMES		
	START	STOP	TOTAL
1.			
2.			
3.			
4.			
5.			
6.			
7.			
TOTAL HOURS FOR SESSION			
TOTAL FORWARD			
TOTAL TO DATE			

FLIGHT MAP

POSTFLIGHT NOTES / JOURNAL ENTRIES:

I CERTIFY THAT THE FOREGOING ENTRIES ARE TRUE AND CORRECT:

PILOT: _____   SPOTTER: _____

droneprep.com

UAS Pilot Log: Unmanned Aircraft Systems Logbook for Drone Pilots and Operators

FLIGHT ID / NO.: _____   Month _____ Day _____ Year _____

FLIGHT LOCATION: _____   Weather: _____

UNMANNED AIRCRAFT SYSTEM: **CREW:**

MANUFACTURER: _____   PILOT: _____

MODEL NUMBER: _____   SPOTTER: _____

PREFLIGHT CHECKLIST:

- ☐ Batteries Charged & Secure
- ☐ Aircraft Hardware OK
- ☐ Equipment & Gear OK
- ☐ Transmitter Controls OK
- ☐ Props OK & Tight
- ☐ Software / Firmware Update
- ☐ Transmitter Control Power ON
- ☐ Aircraft Power ON
- ☐ Compass Calibration
- ☐ Camera / FPV ON
- ☐ Satellite Connection
- ☐ Applications / Systems ON

PREFLIGHT NOTES:

SESSION FLIGHT INTERVALS	FLIGHT TIMES		
	START	STOP	TOTAL
1.			
2.			
3.			
4.			
5.			
6.			
7.			
TOTAL HOURS FOR SESSION			
TOTAL FORWARD			
TOTAL TO DATE			

FLIGHT MAP

POSTFLIGHT NOTES / JOURNAL ENTRIES:

I CERTIFY THAT THE FOREGOING ENTRIES ARE TRUE AND CORRECT:

PILOT: _____   SPOTTER: _____

droneprep.com

UAS Pilot Log: Unmanned Aircraft Systems Logbook for Drone Pilots and Operators

FLIGHT ID / NO.: _____   Month _____ Day _____ Year _____

FLIGHT LOCATION: _____   Weather: _____

UNMANNED AIRCRAFT SYSTEM: **CREW:**

MANUFACTURER: _____   PILOT: _____

MODEL NUMBER: _____   SPOTTER: _____

PREFLIGHT CHECKLIST:
- ☐ Batteries Charged & Secure
- ☐ Aircraft Hardware OK
- ☐ Equipment & Gear OK
- ☐ Transmitter Controls OK
- ☐ Props OK & Tight
- ☐ Software / Firmware Update
- ☐ Transmitter Control Power ON
- ☐ Aircraft Power ON
- ☐ Compass Calibration
- ☐ Camera / FPV ON
- ☐ Satellite Connection
- ☐ Applications / Systems ON

PREFLIGHT NOTES:

SESSION FLIGHT INTERVALS	FLIGHT TIMES		
	START	STOP	TOTAL
1.			
2.			
3.			
4.			
5.			
6.			
7.			
TOTAL HOURS FOR SESSION			
TOTAL FORWARD			
TOTAL TO DATE			

FLIGHT MAP

POSTFLIGHT NOTES / JOURNAL ENTRIES:

I CERTIFY THAT THE FOREGOING ENTRIES ARE TRUE AND CORRECT:

PILOT: _____   SPOTTER: _____

droneprep.com

UAS Pilot Log: Unmanned Aircraft Systems Logbook for Drone Pilots and Operators

FLIGHT ID / NO.: _____ Month _____ Day _____ Year _____

FLIGHT LOCATION: _____ Weather: _____

UNMANNED AIRCRAFT SYSTEM: **CREW:**

MANUFACTURER: _____   PILOT: _____

MODEL NUMBER: _____   SPOTTER: _____

PREFLIGHT CHECKLIST:
- ☐ Batteries Charged & Secure
- ☐ Aircraft Hardware OK
- ☐ Equipment & Gear OK
- ☐ Transmitter Controls OK
- ☐ Props OK & Tight
- ☐ Software / Firmware Update
- ☐ Transmitter Control Power ON
- ☐ Aircraft Power ON
- ☐ Compass Calibration
- ☐ Camera / FPV ON
- ☐ Satellite Connection
- ☐ Applications / Systems ON

PREFLIGHT NOTES:

SESSION FLIGHT INTERVALS	FLIGHT TIMES		
	START	STOP	TOTAL
1.			
2.			
3.			
4.			
5.			
6.			
7.			
TOTAL HOURS FOR SESSION			
TOTAL FORWARD			
TOTAL TO DATE			

FLIGHT MAP

POSTFLIGHT NOTES / JOURNAL ENTRIES:

I CERTIFY THAT THE FOREGOING ENTRIES ARE TRUE AND CORRECT:

PILOT: _____   SPOTTER: _____

droneprep.com

UAS Pilot Log: Unmanned Aircraft Systems Logbook for Drone Pilots and Operators

FLIGHT ID / NO.: _____

Month _____ Day _____ Year _____

FLIGHT LOCATION: _____

Weather: _____

UNMANNED AIRCRAFT SYSTEM:

CREW:

MANUFACTURER: _____

PILOT: _____

MODEL NUMBER: _____

SPOTTER: _____

PREFLIGHT CHECKLIST:

- ☐ Batteries Charged & Secure
- ☐ Aircraft Hardware OK
- ☐ Equipment & Gear OK
- ☐ Transmitter Controls OK

- ☐ Props OK & Tight
- ☐ Software / Firmware Update
- ☐ Transmitter Control Power ON
- ☐ Aircraft Power ON

- ☐ Compass Calibration
- ☐ Camera / FPV ON
- ☐ Satellite Connection
- ☐ Applications / Systems ON

PREFLIGHT NOTES:

SESSION FLIGHT INTERVALS	FLIGHT TIMES		
	START	STOP	TOTAL
1.			
2.			
3.			
4.			
5.			
6.			
7.			
TOTAL HOURS FOR SESSION			
TOTAL FORWARD			
TOTAL TO DATE			

FLIGHT MAP

POSTFLIGHT NOTES / JOURNAL ENTRIES:

I CERTIFY THAT THE FOREGOING ENTRIES ARE TRUE AND CORRECT:

PILOT: _____

SPOTTER: _____

UAS Pilot Log: Unmanned Aircraft Systems Logbook for Drone Pilots and Operators

FLIGHT ID / NO.: _____ Month _____ Day _____ Year _____

FLIGHT LOCATION: _____ Weather: _____

UNMANNED AIRCRAFT SYSTEM: **CREW:**

MANUFACTURER: _____ PILOT: _____

MODEL NUMBER: _____ SPOTTER: _____

PREFLIGHT CHECKLIST:

- ☐ Batteries Charged & Secure
- ☐ Aircraft Hardware OK
- ☐ Equipment & Gear OK
- ☐ Transmitter Controls OK
- ☐ Props OK & Tight
- ☐ Software / Firmware Update
- ☐ Transmitter Control Power ON
- ☐ Aircraft Power ON
- ☐ Compass Calibration
- ☐ Camera / FPV ON
- ☐ Satellite Connection
- ☐ Applications / Systems ON

PREFLIGHT NOTES:

SESSION FLIGHT INTERVALS	FLIGHT TIMES		
	START	STOP	TOTAL
1.			
2.			
3.			
4.			
5.			
6.			
7.			
TOTAL HOURS FOR SESSION			
TOTAL FORWARD			
TOTAL TO DATE			

FLIGHT MAP

POSTFLIGHT NOTES / JOURNAL ENTRIES:

I CERTIFY THAT THE FOREGOING ENTRIES ARE TRUE AND CORRECT:

PILOT: _____ SPOTTER: _____

droneprep.com

UAS Pilot Log: Unmanned Aircraft Systems Logbook for Drone Pilots and Operators

FLIGHT ID / NO.: _____

Month _____ Day _____ Year _____

FLIGHT LOCATION: _____

Weather: _____

UNMANNED AIRCRAFT SYSTEM:

MANUFACTURER: _____

MODEL NUMBER: _____

CREW:

PILOT: _____

SPOTTER: _____

PREFLIGHT CHECKLIST:

- ☐ Batteries Charged & Secure
- ☐ Aircraft Hardware OK
- ☐ Equipment & Gear OK
- ☐ Transmitter Controls OK
- ☐ Props OK & Tight
- ☐ Software / Firmware Update
- ☐ Transmitter Control Power ON
- ☐ Aircraft Power ON
- ☐ Compass Calibration
- ☐ Camera / FPV ON
- ☐ Satellite Connection
- ☐ Applications / Systems ON

PREFLIGHT NOTES:

_____

SESSION FLIGHT INTERVALS	FLIGHT TIMES		
	START	STOP	TOTAL
1.			
2.			
3.			
4.			
5.			
6.			
7.			
TOTAL HOURS FOR SESSION			
TOTAL FORWARD			
TOTAL TO DATE			

FLIGHT MAP

POSTFLIGHT NOTES / JOURNAL ENTRIES:

I CERTIFY THAT THE FOREGOING ENTRIES ARE TRUE AND CORRECT:

PILOT: _____   SPOTTER: _____

droneprep.com

UAS Pilot Log: Unmanned Aircraft Systems Logbook for Drone Pilots and Operators

FLIGHT ID / NO.: _____   Month _____ Day _____ Year _____

FLIGHT LOCATION: _____   Weather: _____

UNMANNED AIRCRAFT SYSTEM: CREW:

MANUFACTURER: _____   PILOT: _____

MODEL NUMBER: _____   SPOTTER: _____

PREFLIGHT CHECKLIST:

- ☐ Batteries Charged & Secure
- ☐ Aircraft Hardware OK
- ☐ Equipment & Gear OK
- ☐ Transmitter Controls OK

- ☐ Props OK & Tight
- ☐ Software / Firmware Update
- ☐ Transmitter Control Power ON
- ☐ Aircraft Power ON

- ☐ Compass Calibration
- ☐ Camera / FPV ON
- ☐ Satellite Connection
- ☐ Applications / Systems ON

PREFLIGHT NOTES:

_____

SESSION FLIGHT INTERVALS	FLIGHT TIMES		
	START	STOP	TOTAL
1.			
2.			
3.			
4.			
5.			
6.			
7.			
TOTAL HOURS FOR SESSION			
TOTAL FORWARD			
TOTAL TO DATE			

FLIGHT MAP

POSTFLIGHT NOTES / JOURNAL ENTRIES:

I CERTIFY THAT THE FOREGOING ENTRIES ARE TRUE AND CORRECT:

PILOT: _____   SPOTTER: _____

UAS Pilot Log: Unmanned Aircraft Systems Logbook for Drone Pilots and Operators

FLIGHT ID / NO.: _____

Month _____ Day _____ Year _____

FLIGHT LOCATION: _____

Weather: _____

UNMANNED AIRCRAFT SYSTEM:

CREW:

MANUFACTURER: _____

PILOT: _____

MODEL NUMBER: _____

SPOTTER: _____

PREFLIGHT CHECKLIST:
- ☐ Batteries Charged & Secure
- ☐ Aircraft Hardware OK
- ☐ Equipment & Gear OK
- ☐ Transmitter Controls OK
- ☐ Props OK & Tight
- ☐ Software / Firmware Update
- ☐ Transmitter Control Power ON
- ☐ Aircraft Power ON
- ☐ Compass Calibration
- ☐ Camera / FPV ON
- ☐ Satellite Connection
- ☐ Applications / Systems ON

PREFLIGHT NOTES:

SESSION FLIGHT INTERVALS	FLIGHT TIMES		
	START	STOP	TOTAL
1.			
2.			
3.			
4.			
5.			
6.			
7.			
TOTAL HOURS FOR SESSION			
TOTAL FORWARD			
TOTAL TO DATE			

FLIGHT MAP

POSTFLIGHT NOTES / JOURNAL ENTRIES:

I CERTIFY THAT THE FOREGOING ENTRIES ARE TRUE AND CORRECT:

PILOT: _____   SPOTTER: _____

droneprep.com

UAS Pilot Log: Unmanned Aircraft Systems Logbook for Drone Pilots and Operators

FLIGHT ID / NO.: _____

Month _____ Day _____ Year _____

FLIGHT LOCATION: _____

Weather: _____

UNMANNED AIRCRAFT SYSTEM:

CREW:

MANUFACTURER: _____

PILOT: _____

MODEL NUMBER: _____

SPOTTER: _____

PREFLIGHT CHECKLIST:

- ☐ Batteries Charged & Secure
- ☐ Aircraft Hardware OK
- ☐ Equipment & Gear OK
- ☐ Transmitter Controls OK
- ☐ Props OK & Tight
- ☐ Software / Firmware Update
- ☐ Transmitter Control Power ON
- ☐ Aircraft Power ON
- ☐ Compass Calibration
- ☐ Camera / FPV ON
- ☐ Satellite Connection
- ☐ Applications / Systems ON

PREFLIGHT NOTES:

_____

SESSION FLIGHT INTERVALS	FLIGHT TIMES		
	START	STOP	TOTAL
1.			
2.			
3.			
4.			
5.			
6.			
7.			
TOTAL HOURS FOR SESSION			
TOTAL FORWARD			
TOTAL TO DATE			

FLIGHT MAP

POSTFLIGHT NOTES / JOURNAL ENTRIES:

I CERTIFY THAT THE FOREGOING ENTRIES ARE TRUE AND CORRECT:

PILOT: _____

SPOTTER: _____

UAS Pilot Log: Unmanned Aircraft Systems Logbook for Drone Pilots and Operators

FLIGHT ID / NO.: _____   Month _____ Day _____ Year _____

FLIGHT LOCATION: _____   Weather: _____

UNMANNED AIRCRAFT SYSTEM: **CREW:**

MANUFACTURER: _____   PILOT: _____

MODEL NUMBER: _____   SPOTTER: _____

PREFLIGHT CHECKLIST:

- ☐ Batteries Charged & Secure   ☐ Props OK & Tight   ☐ Compass Calibration
- ☐ Aircraft Hardware OK   ☐ Software / Firmware Update   ☐ Camera / FPV ON
- ☐ Equipment & Gear OK   ☐ Transmitter Control Power ON   ☐ Satellite Connection
- ☐ Transmitter Controls OK   ☐ Aircraft Power ON   ☐ Applications / Systems ON

PREFLIGHT NOTES:

SESSION FLIGHT INTERVALS	FLIGHT TIMES		
	START	STOP	TOTAL
1.			
2.			
3.			
4.			
5.			
6.			
7.			
TOTAL HOURS FOR SESSION			
TOTAL FORWARD			
TOTAL TO DATE			

FLIGHT MAP

POSTFLIGHT NOTES / JOURNAL ENTRIES:

I CERTIFY THAT THE FOREGOING ENTRIES ARE TRUE AND CORRECT:

PILOT: _____   SPOTTER: _____

UAS Pilot Log: Unmanned Aircraft Systems Logbook for Drone Pilots and Operators

FLIGHT ID / NO.: _____   Month _____ Day _____ Year _____

FLIGHT LOCATION: _____   Weather: _____

UNMANNED AIRCRAFT SYSTEM: **CREW:**

MANUFACTURER: _____   PILOT: _____

MODEL NUMBER: _____   SPOTTER: _____

PREFLIGHT CHECKLIST:

- ☐ Batteries Charged & Secure
- ☐ Aircraft Hardware OK
- ☐ Equipment & Gear OK
- ☐ Transmitter Controls OK
- ☐ Props OK & Tight
- ☐ Software / Firmware Update
- ☐ Transmitter Control Power ON
- ☐ Aircraft Power ON
- ☐ Compass Calibration
- ☐ Camera / FPV ON
- ☐ Satellite Connection
- ☐ Applications / Systems ON

PREFLIGHT NOTES:

_____

SESSION FLIGHT INTERVALS	FLIGHT TIMES		
	START	STOP	TOTAL
1.			
2.			
3.			
4.			
5.			
6.			
7.			
TOTAL HOURS FOR SESSION			
TOTAL FORWARD			
TOTAL TO DATE			

FLIGHT MAP

POSTFLIGHT NOTES / JOURNAL ENTRIES:

I CERTIFY THAT THE FOREGOING ENTRIES ARE TRUE AND CORRECT:

PILOT: _____   SPOTTER: _____

UAS Pilot Log: Unmanned Aircraft Systems Logbook for Drone Pilots and Operators

FLIGHT ID / NO.: _____   Month _____ Day _____ Year _____

FLIGHT LOCATION: _____   Weather: _____

UNMANNED AIRCRAFT SYSTEM: **CREW:**

MANUFACTURER: _____   PILOT: _____

MODEL NUMBER: _____   SPOTTER: _____

PREFLIGHT CHECKLIST:
- ☐ Batteries Charged & Secure   ☐ Props OK & Tight   ☐ Compass Calibration
- ☐ Aircraft Hardware OK   ☐ Software / Firmware Update   ☐ Camera / FPV ON
- ☐ Equipment & Gear OK   ☐ Transmitter Control Power ON   ☐ Satellite Connection
- ☐ Transmitter Controls OK   ☐ Aircraft Power ON   ☐ Applications / Systems ON

PREFLIGHT NOTES:

_____

SESSION FLIGHT INTERVALS	FLIGHT TIMES		
	START	STOP	TOTAL
1.			
2.			
3.			
4.			
5.			
6.			
7.			
TOTAL HOURS FOR SESSION			
TOTAL FORWARD			
TOTAL TO DATE			

FLIGHT MAP

POSTFLIGHT NOTES / JOURNAL ENTRIES:

I CERTIFY THAT THE FOREGOING ENTRIES ARE TRUE AND CORRECT:

PILOT: _____   SPOTTER: _____

UAS Pilot Log: Unmanned Aircraft Systems Logbook for Drone Pilots and Operators

FLIGHT ID / NO.: _____   Month _____ Day _____ Year _____

FLIGHT LOCATION: _____   Weather: _____

UNMANNED AIRCRAFT SYSTEM: **CREW:**

MANUFACTURER: _____   PILOT: _____

MODEL NUMBER: _____   SPOTTER: _____

PREFLIGHT CHECKLIST:

- ☐ Batteries Charged & Secure
- ☐ Aircraft Hardware OK
- ☐ Equipment & Gear OK
- ☐ Transmitter Controls OK
- ☐ Props OK & Tight
- ☐ Software / Firmware Update
- ☐ Transmitter Control Power ON
- ☐ Aircraft Power ON
- ☐ Compass Calibration
- ☐ Camera / FPV ON
- ☐ Satellite Connection
- ☐ Applications / Systems ON

PREFLIGHT NOTES:

_____

SESSION FLIGHT INTERVALS	FLIGHT TIMES		
	START	STOP	TOTAL
1.			
2.			
3.			
4.			
5.			
6.			
7.			
TOTAL HOURS FOR SESSION			
TOTAL FORWARD			
TOTAL TO DATE			

FLIGHT MAP

POSTFLIGHT NOTES / JOURNAL ENTRIES:

I CERTIFY THAT THE FOREGOING ENTRIES ARE TRUE AND CORRECT:

PILOT: _____   SPOTTER: _____

UAS Pilot Log: Unmanned Aircraft Systems Logbook for Drone Pilots and Operators

FLIGHT ID / NO.: _____ Month _____ Day _____ Year _____

FLIGHT LOCATION: _____ Weather: _____

UNMANNED AIRCRAFT SYSTEM: **CREW:**

MANUFACTURER: _____   PILOT: _____

MODEL NUMBER: _____   SPOTTER: _____

PREFLIGHT CHECKLIST:
- ☐ Batteries Charged & Secure
- ☐ Aircraft Hardware OK
- ☐ Equipment & Gear OK
- ☐ Transmitter Controls OK
- ☐ Props OK & Tight
- ☐ Software / Firmware Update
- ☐ Transmitter Control Power ON
- ☐ Aircraft Power ON
- ☐ Compass Calibration
- ☐ Camera / FPV ON
- ☐ Satellite Connection
- ☐ Applications / Systems ON

PREFLIGHT NOTES:

SESSION FLIGHT INTERVALS	FLIGHT TIMES		
	START	STOP	TOTAL
1.			
2.			
3.			
4.			
5.			
6.			
7.			
TOTAL HOURS FOR SESSION			
TOTAL FORWARD			
TOTAL TO DATE			

FLIGHT MAP

POSTFLIGHT NOTES / JOURNAL ENTRIES:

I CERTIFY THAT THE FOREGOING ENTRIES ARE TRUE AND CORRECT:

PILOT: _____   SPOTTER: _____

droneprep.com

UAS Pilot Log: Unmanned Aircraft Systems Logbook for Drone Pilots and Operators

FLIGHT ID / NO.: _____

Month _____ Day _____ Year _____

FLIGHT LOCATION: _____

Weather: _____

UNMANNED AIRCRAFT SYSTEM:

CREW:

MANUFACTURER: _____

PILOT: _____

MODEL NUMBER: _____

SPOTTER: _____

PREFLIGHT CHECKLIST:

- ☐ Batteries Charged & Secure
- ☐ Aircraft Hardware OK
- ☐ Equipment & Gear OK
- ☐ Transmitter Controls OK

- ☐ Props OK & Tight
- ☐ Software / Firmware Update
- ☐ Transmitter Control Power ON
- ☐ Aircraft Power ON

- ☐ Compass Calibration
- ☐ Camera / FPV ON
- ☐ Satellite Connection
- ☐ Applications / Systems ON

PREFLIGHT NOTES:

SESSION FLIGHT INTERVALS	FLIGHT TIMES		
	START	STOP	TOTAL
1.			
2.			
3.			
4.			
5.			
6.			
7.			
TOTAL HOURS FOR SESSION			
TOTAL FORWARD			
TOTAL TO DATE			

FLIGHT MAP

POSTFLIGHT NOTES / JOURNAL ENTRIES:

I CERTIFY THAT THE FOREGOING ENTRIES ARE TRUE AND CORRECT:

PILOT: _____

SPOTTER: _____

UAS Pilot Log: Unmanned Aircraft Systems Logbook for Drone Pilots and Operators

FLIGHT ID / NO.: _____ Month _____ Day _____ Year _____

FLIGHT LOCATION: _____ Weather: _____

UNMANNED AIRCRAFT SYSTEM: **CREW:**

MANUFACTURER: _____ PILOT: _____

MODEL NUMBER: _____ SPOTTER: _____

PREFLIGHT CHECKLIST:
- ☐ Batteries Charged & Secure
- ☐ Aircraft Hardware OK
- ☐ Equipment & Gear OK
- ☐ Transmitter Controls OK
- ☐ Props OK & Tight
- ☐ Software / Firmware Update
- ☐ Transmitter Control Power ON
- ☐ Aircraft Power ON
- ☐ Compass Calibration
- ☐ Camera / FPV ON
- ☐ Satellite Connection
- ☐ Applications / Systems ON

PREFLIGHT NOTES:

SESSION FLIGHT INTERVALS	FLIGHT TIMES		
	START	STOP	TOTAL
1.			
2.			
3.			
4.			
5.			
6.			
7.			
TOTAL HOURS FOR SESSION			
TOTAL FORWARD			
TOTAL TO DATE			

FLIGHT MAP

POSTFLIGHT NOTES / JOURNAL ENTRIES:

I CERTIFY THAT THE FOREGOING ENTRIES ARE TRUE AND CORRECT:

PILOT: _____ SPOTTER: _____

droneprep.com

UAS Pilot Log: Unmanned Aircraft Systems Logbook for Drone Pilots and Operators

FLIGHT ID / NO.: _____ Month _____ Day _____ Year _____

FLIGHT LOCATION: _____ Weather: _____

UNMANNED AIRCRAFT SYSTEM: **CREW:**

MANUFACTURER: _____   PILOT: _____

MODEL NUMBER: _____   SPOTTER: _____

PREFLIGHT CHECKLIST:

- ☐ Batteries Charged & Secure
- ☐ Aircraft Hardware OK
- ☐ Equipment & Gear OK
- ☐ Transmitter Controls OK
- ☐ Props OK & Tight
- ☐ Software / Firmware Update
- ☐ Transmitter Control Power ON
- ☐ Aircraft Power ON
- ☐ Compass Calibration
- ☐ Camera / FPV ON
- ☐ Satellite Connection
- ☐ Applications / Systems ON

PREFLIGHT NOTES:

_____

SESSION FLIGHT INTERVALS	FLIGHT TIMES		
	START	STOP	TOTAL
1.			
2.			
3.			
4.			
5.			
6.			
7.			
TOTAL HOURS FOR SESSION			
TOTAL FORWARD			
TOTAL TO DATE			

FLIGHT MAP

POSTFLIGHT NOTES / JOURNAL ENTRIES:

I CERTIFY THAT THE FOREGOING ENTRIES ARE TRUE AND CORRECT:

PILOT: _____   SPOTTER: _____

droneprep.com

UAS Pilot Log: Unmanned Aircraft Systems Logbook for Drone Pilots and Operators

FLIGHT ID / NO.: _____

Month _____ Day _____ Year _____

FLIGHT LOCATION: _____

Weather: _____

UNMANNED AIRCRAFT SYSTEM:

CREW:

MANUFACTURER: _____

PILOT: _____

MODEL NUMBER: _____

SPOTTER: _____

PREFLIGHT CHECKLIST:
- ☐ Batteries Charged & Secure
- ☐ Aircraft Hardware OK
- ☐ Equipment & Gear OK
- ☐ Transmitter Controls OK
- ☐ Props OK & Tight
- ☐ Software / Firmware Update
- ☐ Transmitter Control Power ON
- ☐ Aircraft Power ON
- ☐ Compass Calibration
- ☐ Camera / FPV ON
- ☐ Satellite Connection
- ☐ Applications / Systems ON

PREFLIGHT NOTES:

SESSION FLIGHT INTERVALS	FLIGHT TIMES		
	START	STOP	TOTAL
1.			
2.			
3.			
4.			
5.			
6.			
7.			
TOTAL HOURS FOR SESSION			
TOTAL FORWARD			
TOTAL TO DATE			

FLIGHT MAP

POSTFLIGHT NOTES / JOURNAL ENTRIES:

I CERTIFY THAT THE FOREGOING ENTRIES ARE TRUE AND CORRECT:

PILOT: _____   SPOTTER: _____

UAS Pilot Log: Unmanned Aircraft Systems Logbook for Drone Pilots and Operators

FLIGHT ID / NO.: _____   Month _____ Day _____ Year _____

FLIGHT LOCATION: _____   Weather: _____

UNMANNED AIRCRAFT SYSTEM: **CREW:**

MANUFACTURER: _____   PILOT: _____

MODEL NUMBER: _____   SPOTTER: _____

PREFLIGHT CHECKLIST:

- ☐ Batteries Charged & Secure
- ☐ Aircraft Hardware OK
- ☐ Equipment & Gear OK
- ☐ Transmitter Controls OK
- ☐ Props OK & Tight
- ☐ Software / Firmware Update
- ☐ Transmitter Control Power ON
- ☐ Aircraft Power ON
- ☐ Compass Calibration
- ☐ Camera / FPV ON
- ☐ Satellite Connection
- ☐ Applications / Systems ON

PREFLIGHT NOTES:

_____

SESSION FLIGHT INTERVALS	FLIGHT TIMES		
	START	STOP	TOTAL
1.			
2.			
3.			
4.			
5.			
6.			
7.			
TOTAL HOURS FOR SESSION			
TOTAL FORWARD			
TOTAL TO DATE			

FLIGHT MAP

POSTFLIGHT NOTES / JOURNAL ENTRIES:

I CERTIFY THAT THE FOREGOING ENTRIES ARE TRUE AND CORRECT:

PILOT: _____   SPOTTER: _____

UAS Pilot Log: Unmanned Aircraft Systems Logbook for Drone Pilots and Operators

FLIGHT ID / NO.: _____

Month _____ Day _____ Year _____

FLIGHT LOCATION: _____

Weather: _____

UNMANNED AIRCRAFT SYSTEM:

CREW:

MANUFACTURER: _____

PILOT: _____

MODEL NUMBER: _____

SPOTTER: _____

PREFLIGHT CHECKLIST:
- ☐ Batteries Charged & Secure
- ☐ Aircraft Hardware OK
- ☐ Equipment & Gear OK
- ☐ Transmitter Controls OK
- ☐ Props OK & Tight
- ☐ Software / Firmware Update
- ☐ Transmitter Control Power ON
- ☐ Aircraft Power ON
- ☐ Compass Calibration
- ☐ Camera / FPV ON
- ☐ Satellite Connection
- ☐ Applications / Systems ON

PREFLIGHT NOTES:

SESSION FLIGHT INTERVALS	FLIGHT TIMES		
	START	STOP	TOTAL
1.			
2.			
3.			
4.			
5.			
6.			
7.			
TOTAL HOURS FOR SESSION			
TOTAL FORWARD			
TOTAL TO DATE			

FLIGHT MAP

POSTFLIGHT NOTES / JOURNAL ENTRIES:

I CERTIFY THAT THE FOREGOING ENTRIES ARE TRUE AND CORRECT:

PILOT: _____ SPOTTER: _____

droneprep.com

UAS Pilot Log: Unmanned Aircraft Systems Logbook for Drone Pilots and Operators

FLIGHT ID / NO.: _____ Month _____ Day _____ Year _____

FLIGHT LOCATION: _____ Weather: _____

UNMANNED AIRCRAFT SYSTEM: **CREW:**

MANUFACTURER: _____ PILOT: _____

MODEL NUMBER: _____ SPOTTER: _____

PREFLIGHT CHECKLIST:

- ☐ Batteries Charged & Secure
- ☐ Aircraft Hardware OK
- ☐ Equipment & Gear OK
- ☐ Transmitter Controls OK

- ☐ Props OK & Tight
- ☐ Software / Firmware Update
- ☐ Transmitter Control Power ON
- ☐ Aircraft Power ON

- ☐ Compass Calibration
- ☐ Camera / FPV ON
- ☐ Satellite Connection
- ☐ Applications / Systems ON

PREFLIGHT NOTES:

SESSION FLIGHT INTERVALS	FLIGHT TIMES		
	START	STOP	TOTAL
1.			
2.			
3.			
4.			
5.			
6.			
7.			
TOTAL HOURS FOR SESSION			
TOTAL FORWARD			
TOTAL TO DATE			

FLIGHT MAP

POSTFLIGHT NOTES / JOURNAL ENTRIES:

I CERTIFY THAT THE FOREGOING ENTRIES ARE TRUE AND CORRECT:

PILOT: _____ SPOTTER: _____

UAS Pilot Log: Unmanned Aircraft Systems Logbook for Drone Pilots and Operators

FLIGHT ID / NO.: _____

Month _____ Day _____ Year _____

FLIGHT LOCATION: _____

Weather: _____

UNMANNED AIRCRAFT SYSTEM:

CREW:

MANUFACTURER: _____

PILOT: _____

MODEL NUMBER: _____

SPOTTER: _____

PREFLIGHT CHECKLIST:

- ☐ Batteries Charged & Secure
- ☐ Aircraft Hardware OK
- ☐ Equipment & Gear OK
- ☐ Transmitter Controls OK
- ☐ Props OK & Tight
- ☐ Software / Firmware Update
- ☐ Transmitter Control Power ON
- ☐ Aircraft Power ON
- ☐ Compass Calibration
- ☐ Camera / FPV ON
- ☐ Satellite Connection
- ☐ Applications / Systems ON

PREFLIGHT NOTES:

SESSION FLIGHT INTERVALS	FLIGHT TIMES		
	START	STOP	TOTAL
1.			
2.			
3.			
4.			
5.			
6.			
7.			
TOTAL HOURS FOR SESSION			
TOTAL FORWARD			
TOTAL TO DATE			

FLIGHT MAP

POSTFLIGHT NOTES / JOURNAL ENTRIES:

I CERTIFY THAT THE FOREGOING ENTRIES ARE TRUE AND CORRECT:

PILOT: _____   SPOTTER: _____

UAS Pilot Log: Unmanned Aircraft Systems Logbook for Drone Pilots and Operators

FLIGHT ID / NO.: _____

Month _____ Day _____ Year _____

FLIGHT LOCATION: _____

Weather: _____

UNMANNED AIRCRAFT SYSTEM:

CREW:

MANUFACTURER: _____

PILOT: _____

MODEL NUMBER: _____

SPOTTER: _____

PREFLIGHT CHECKLIST:

- ☐ Batteries Charged & Secure
- ☐ Aircraft Hardware OK
- ☐ Equipment & Gear OK
- ☐ Transmitter Controls OK
- ☐ Props OK & Tight
- ☐ Software / Firmware Update
- ☐ Transmitter Control Power ON
- ☐ Aircraft Power ON
- ☐ Compass Calibration
- ☐ Camera / FPV ON
- ☐ Satellite Connection
- ☐ Applications / Systems ON

PREFLIGHT NOTES:

SESSION FLIGHT INTERVALS	FLIGHT TIMES		
	START	STOP	TOTAL
1.			
2.			
3.			
4.			
5.			
6.			
7.			
TOTAL HOURS FOR SESSION			
TOTAL FORWARD			
TOTAL TO DATE			

FLIGHT MAP

POSTFLIGHT NOTES / JOURNAL ENTRIES:

I CERTIFY THAT THE FOREGOING ENTRIES ARE TRUE AND CORRECT:

PILOT: _____

SPOTTER: _____

droneprep.com

UAS Pilot Log: Unmanned Aircraft Systems Logbook for Drone Pilots and Operators

FLIGHT ID / NO.: _____

Month _____ Day _____ Year _____

FLIGHT LOCATION: _____

Weather: _____

UNMANNED AIRCRAFT SYSTEM:

CREW:

MANUFACTURER: _____

PILOT: _____

MODEL NUMBER: _____

SPOTTER: _____

PREFLIGHT CHECKLIST:

- ☐ Batteries Charged & Secure
- ☐ Aircraft Hardware OK
- ☐ Equipment & Gear OK
- ☐ Transmitter Controls OK
- ☐ Props OK & Tight
- ☐ Software / Firmware Update
- ☐ Transmitter Control Power ON
- ☐ Aircraft Power ON
- ☐ Compass Calibration
- ☐ Camera / FPV ON
- ☐ Satellite Connection
- ☐ Applications / Systems ON

PREFLIGHT NOTES:

SESSION FLIGHT INTERVALS	FLIGHT TIMES		
	START	STOP	TOTAL
1.			
2.			
3.			
4.			
5.			
6.			
7.			
TOTAL HOURS FOR SESSION			
TOTAL FORWARD			
TOTAL TO DATE			

FLIGHT MAP

POSTFLIGHT NOTES / JOURNAL ENTRIES:

I CERTIFY THAT THE FOREGOING ENTRIES ARE TRUE AND CORRECT:

PILOT: _____   SPOTTER: _____

droneprep.com

UAS Pilot Log: Unmanned Aircraft Systems Logbook for Drone Pilots and Operators

FLIGHT ID / NO.: _____

Month _____ Day _____ Year _____

FLIGHT LOCATION: _____

Weather: _____

UNMANNED AIRCRAFT SYSTEM:

CREW:

MANUFACTURER: _____

PILOT: _____

MODEL NUMBER: _____

SPOTTER: _____

PREFLIGHT CHECKLIST:
- ☐ Batteries Charged & Secure
- ☐ Aircraft Hardware OK
- ☐ Equipment & Gear OK
- ☐ Transmitter Controls OK
- ☐ Props OK & Tight
- ☐ Software / Firmware Update
- ☐ Transmitter Control Power ON
- ☐ Aircraft Power ON
- ☐ Compass Calibration
- ☐ Camera / FPV ON
- ☐ Satellite Connection
- ☐ Applications / Systems ON

PREFLIGHT NOTES:

_____

SESSION FLIGHT INTERVALS	FLIGHT TIMES		
	START	STOP	TOTAL
1.			
2.			
3.			
4.			
5.			
6.			
7.			
TOTAL HOURS FOR SESSION			
TOTAL FORWARD			
TOTAL TO DATE			

FLIGHT MAP

POSTFLIGHT NOTES / JOURNAL ENTRIES:

I CERTIFY THAT THE FOREGOING ENTRIES ARE TRUE AND CORRECT:

PILOT: _____ SPOTTER: _____

droneprep.com

UAS Pilot Log: Unmanned Aircraft Systems Logbook for Drone Pilots and Operators

FLIGHT ID / NO.: _____   Month _____ Day _____ Year _____

FLIGHT LOCATION: _____   Weather: _____

UNMANNED AIRCRAFT SYSTEM: **CREW:**

MANUFACTURER: _____   PILOT: _____

MODEL NUMBER: _____   SPOTTER: _____

PREFLIGHT CHECKLIST:
- ☐ Batteries Charged & Secure   ☐ Props OK & Tight   ☐ Compass Calibration
- ☐ Aircraft Hardware OK   ☐ Software / Firmware Update   ☐ Camera / FPV ON
- ☐ Equipment & Gear OK   ☐ Transmitter Control Power ON   ☐ Satellite Connection
- ☐ Transmitter Controls OK   ☐ Aircraft Power ON   ☐ Applications / Systems ON

PREFLIGHT NOTES:

SESSION FLIGHT INTERVALS	FLIGHT TIMES		
	START	STOP	TOTAL
1.			
2.			
3.			
4.			
5.			
6.			
7.			
TOTAL HOURS FOR SESSION			
TOTAL FORWARD			
TOTAL TO DATE			

FLIGHT MAP

POSTFLIGHT NOTES / JOURNAL ENTRIES:

I CERTIFY THAT THE FOREGOING ENTRIES ARE TRUE AND CORRECT:

PILOT: _____   SPOTTER: _____

droneprep.com

UAS Pilot Log: Unmanned Aircraft Systems Logbook for Drone Pilots and Operators

FLIGHT ID / NO.: _____   Month _____ Day _____ Year _____

FLIGHT LOCATION: _____   Weather: _____

UNMANNED AIRCRAFT SYSTEM: **CREW:**

MANUFACTURER: _____   PILOT: _____

MODEL NUMBER: _____   SPOTTER: _____

PREFLIGHT CHECKLIST:

- ☐ Batteries Charged & Secure
- ☐ Aircraft Hardware OK
- ☐ Equipment & Gear OK
- ☐ Transmitter Controls OK
- ☐ Props OK & Tight
- ☐ Software / Firmware Update
- ☐ Transmitter Control Power ON
- ☐ Aircraft Power ON
- ☐ Compass Calibration
- ☐ Camera / FPV ON
- ☐ Satellite Connection
- ☐ Applications / Systems ON

PREFLIGHT NOTES:

SESSION FLIGHT INTERVALS	FLIGHT TIMES		
	START	STOP	TOTAL
1.			
2.			
3.			
4.			
5.			
6.			
7.			
TOTAL HOURS FOR SESSION			
TOTAL FORWARD			
TOTAL TO DATE			

FLIGHT MAP

POSTFLIGHT NOTES / JOURNAL ENTRIES:

I CERTIFY THAT THE FOREGOING ENTRIES ARE TRUE AND CORRECT:

PILOT: _____   SPOTTER: _____

UAS Pilot Log: Unmanned Aircraft Systems Logbook for Drone Pilots and Operators

FLIGHT ID / NO.: _____

Month _____ Day _____ Year _____

FLIGHT LOCATION: _____

Weather: _____

UNMANNED AIRCRAFT SYSTEM:

CREW:

MANUFACTURER: _____

PILOT: _____

MODEL NUMBER: _____

SPOTTER: _____

PREFLIGHT CHECKLIST:
- ☐ Batteries Charged & Secure
- ☐ Aircraft Hardware OK
- ☐ Equipment & Gear OK
- ☐ Transmitter Controls OK
- ☐ Props OK & Tight
- ☐ Software / Firmware Update
- ☐ Transmitter Control Power ON
- ☐ Aircraft Power ON
- ☐ Compass Calibration
- ☐ Camera / FPV ON
- ☐ Satellite Connection
- ☐ Applications / Systems ON

PREFLIGHT NOTES:

SESSION FLIGHT INTERVALS	FLIGHT TIMES		
	START	STOP	TOTAL
1.			
2.			
3.			
4.			
5.			
6.			
7.			
TOTAL HOURS FOR SESSION			
TOTAL FORWARD			
TOTAL TO DATE			

FLIGHT MAP

POSTFLIGHT NOTES / JOURNAL ENTRIES:

I CERTIFY THAT THE FOREGOING ENTRIES ARE TRUE AND CORRECT:

PILOT: _____   SPOTTER: _____

UAS Pilot Log: Unmanned Aircraft Systems Logbook for Drone Pilots and Operators

FLIGHT ID / NO.: _____

Month _____ Day _____ Year _____

FLIGHT LOCATION: _____

Weather: _____

UNMANNED AIRCRAFT SYSTEM:

CREW:

MANUFACTURER: _____

PILOT: _____

MODEL NUMBER: _____

SPOTTER: _____

PREFLIGHT CHECKLIST:

- ☐ Batteries Charged & Secure
- ☐ Aircraft Hardware OK
- ☐ Equipment & Gear OK
- ☐ Transmitter Controls OK
- ☐ Props OK & Tight
- ☐ Software / Firmware Update
- ☐ Transmitter Control Power ON
- ☐ Aircraft Power ON
- ☐ Compass Calibration
- ☐ Camera / FPV ON
- ☐ Satellite Connection
- ☐ Applications / Systems ON

PREFLIGHT NOTES:

SESSION FLIGHT INTERVALS	FLIGHT TIMES		
	START	STOP	TOTAL
1.			
2.			
3.			
4.			
5.			
6.			
7.			
TOTAL HOURS FOR SESSION			
TOTAL FORWARD			
TOTAL TO DATE			

FLIGHT MAP

POSTFLIGHT NOTES / JOURNAL ENTRIES:

I CERTIFY THAT THE FOREGOING ENTRIES ARE TRUE AND CORRECT:

PILOT: _____

SPOTTER: _____

UAS Pilot Log: Unmanned Aircraft Systems Logbook for Drone Pilots and Operators

FLIGHT ID / NO.: _____

Month _____ Day _____ Year _____

FLIGHT LOCATION: _____

Weather: _____

UNMANNED AIRCRAFT SYSTEM:

MANUFACTURER: _____

MODEL NUMBER: _____

CREW:

PILOT: _____

SPOTTER: _____

PREFLIGHT CHECKLIST:
- ☐ Batteries Charged & Secure
- ☐ Aircraft Hardware OK
- ☐ Equipment & Gear OK
- ☐ Transmitter Controls OK
- ☐ Props OK & Tight
- ☐ Software / Firmware Update
- ☐ Transmitter Control Power ON
- ☐ Aircraft Power ON
- ☐ Compass Calibration
- ☐ Camera / FPV ON
- ☐ Satellite Connection
- ☐ Applications / Systems ON

PREFLIGHT NOTES:

SESSION FLIGHT INTERVALS	FLIGHT TIMES		
	START	STOP	TOTAL
1.			
2.			
3.			
4.			
5.			
6.			
7.			
TOTAL HOURS FOR SESSION			
TOTAL FORWARD			
TOTAL TO DATE			

FLIGHT MAP

POSTFLIGHT NOTES / JOURNAL ENTRIES:

I CERTIFY THAT THE FOREGOING ENTRIES ARE TRUE AND CORRECT:

PILOT: _____   SPOTTER: _____

Printed in Germany
by Amazon Distribution
GmbH, Leipzig